Refrigeration Equipment

Refrigeration Equipment

A servicing and installation handbook

Second edition

A. C. Bryant

AMSTERDAM • BOSTON • HEIDELBERG • LONDON • NEW YORK • OXFORD
PARIS • SAN DIEGO • SAN FRANCISCO • SINGAPORE • SYDNEY • TOKYO
Newnes is an imprint of Elsevier

ELSEVIER

Newnes

Newnes
An imprint of Elsevier
Linacre House, Jordan Hill, Oxford OX2 8DP
30 Corporate Drive, Burlington, MA 01803

First published 1991
Reprinted 1993 (twice)
Second edition 1997
Reprinted 1991, 2000 (twice), 2001, 2002, 2003, 2004
Transferred to digital printing 2004, 2005, 2006

British Library Cataloguing in Publication Data
A catalogue record for this book is available from the British Library

Library of Congress Cataloguing in Publication Data
A catalogue record for this book is available from the Library of Congress

ISBN 0 7506 3688 2

For information on all Newnes publications
visit our website at www.newnespress.com

Working together to grow
libraries in developing countries

www.elsevier.com | www.bookaid.org | www.sabre.org

ELSEVIER   BOOK AID
            International   Sabre Foundation

Contents

Preface   ix

Acknowledgements   x

1   Introduction   1
The service engineer   1
Installation procedures   2
Vapour compression systems   2

Part One   Servicing

2   Service valves and gauges   7
System service valves   7
Service gauge manifold   8
Fitting gauges   11
Removing gauges   13
Other valves   14

3   Refrigerant leak detection   17
Dye systems   17
Test pressures   17
Methods of testing   18
Pressure leak testing   20
Handling refrigerants: safety precautions   21

4   Pumping down and charging   24
Pumping down   24
Adding refrigerant   25
Vapour charging   25
Liquid charging   27
Refrigerant cylinders and fill ratio   29

5   Operating principles and symptoms   31
Temperature, humidity and air motion in food storage   31
Calculating the operating head pressure   34
Compressor pressure ratio   37
Symptoms of system faults   38

6   Service diagnosis and repairs   41
Pressure drop: external equalizing   43
Suction line frosting   45
Distributor refrigerant control   46
Replacing a thermostatic expansion valve   46
Replacing the filter drier   49
Low side purging   50
Moisture in the system   50
Compressor efficiency test   51
Removing compressor valve plate assembly   52
Removing a compressor rotary shaft seal   53
Excessive operating head pressure   55
High side purging   57
Water cooled condensers: scale and corrosion   58
Compressor motor burn-out: system flushing   60
Pressure controls   64
Motor cycling controls: thermostats   69
System faults   71
Noise   71
Seven simple steps for service diagnosis   72

7   Electrical equipment and service   74
Description   74
Service equipment   75
Current relay   75
Potential (voltage) relay   77
Solid state starting devices   77
Capacitors   80
Centrifugal switch (open-type motors)   82
Motor protection   83
Electrical test procedures   84

8   Domestic refrigerators and freezers   90
Appliance systems   90
Components and operations   91
Electrical faults   96
Decontaminating domestic systems   97
Refrigerant charging in domestic appliances   97
The domestic absorption system   101

9   Fault finding guide for vapour compression systems   104
Visual fault finding   104

Pressures   105
Advanced diagnosis   106
Single phase/three phase motor compressors and remote-drive
motors   107

Part Two   Installation and Commissioning

10   **Pipework and oil traps   111**
Installation principles   111
Pipework fittings   112
Pipework supports   113
Pipework routes   115
Oil traps   117
Oil separators   118
Discharge line mufflers   120
Parallel pipework   121
Pipework assembly   124

11   **System control valves   128**
Crankcase pressure regulator   128
Evaporator pressure regulator   129
Water regulating valve   129
Reversing valve   131
Fusible plugs   131

12   **Other system components   133**
Solenoid valve   133
Crankcase heater   133
Check valve   134
Sight glass   134
Filter drier   135
Oil pressure failure switch   136

13   **Drive belts and couplings   139**
Drive belts   139
Drive couplings   142

14   **Electrical circuit protection   147**
Fuses   147
Circuit breakers   150

15   **Drive motors   152**
Installation details   152
Three phase electrical connections   155

16   Commissioning of refrigerating systems   159
Contaminants   159
Evacuation   160
Commissioning checks   163
Oil addition and removal   164
Environmental impact of CFCs   169
Good refrigeration practice for CFC systems   170
Refrigerant recovery system   171

Part Three   Replacement Refrigerants and Ozone Depletion

17   Ozone depletion   175
The Montreal Protocol   175

18   New and replacement refrigerants   178
Zeotropic blends   181
Refrigerant blends   181
Reclaiming the refrigerant   185
Changing the refrigerant charge   186
Compressor lubricating oils   187
Domestic and commercial sealed hermetic systems   189

Appendix A   Fundamental principles of air conditioning   194

Appendix B   Refrigerant data   215

Index   219

Preface

This handbook has been written to meet the need for a practical guide to refrigeration systems and their installation, maintenance and repair. It deals with all the basic installation and service procedures and aims to be useful equally to the practising mechanic and to the trainee or student technician.

The first edition of *Refrigeration Equipment* became the standard text for the City & Guilds 2070 course. The new edition will provide underpinning knowledge for the new Level 2 NVQ, and the fundamental for students following Level 3 NVQs, within the City & Guilds 6007 scheme. It also continues to cover for the BTEC National unit in refrigeration technology.

Many refrigeration mechanics receive their training within the industry, either 'on site', or in a manufacturer's in-plant training course. These trainees will find this book a helpful source of information and reference.

Finally, refrigeration installations are to be found in numerous establishments—in the frozen food industry, airports, hotels, hospitals, supermarkets—where maintenance personnel may sometimes be responsible for maintaining an emergency service until specialists are available.

I hope that this handbook will be welcomed by the many services and maintenance engineers whose responsibilities include the monitoring of refrigeration equipment.

A. C. B.

As the Chairman of the Institute of Refrigeration's Management Panel for its Service Engineer Section, I am pleased to recommend Albert Bryant's book, *Refrigeration Equipment*, to all active refrigeration service and maintenance engineers, as it covers a wide range of commercial refrigeration topics, taking into account associated electrical equipment.

Michael Boast, FInstR

Acknowledgements

I wish to acknowledge Danfoss Automatic Controls and Equipment, Carrier Distribution Ltd, ICI Klea, Du Pont-Suva Refrigerants and Ranco Controls Division for providing some of the information in this book.

1 *Introduction*

To be able to install, commission and carry out maintenance or repairs to even the most basic refrigeration system, it is essential that the engineer has a sound knowledge of the system operation, the functions of the various valves, the controls employed in the system, and the specialist tools or equipment necessary to carry out those tasks.

Generally, larger refrigeration companies employ engineers who specialize in servicing and others involved only with the installation of equipment. Smaller companies require engineers to be proficient in both service and installation practice.

This manual deals with the fundamental principles of service, installation and commissioning, but not necessarily to any specific manufacturer's recommendation or equipment.

The service engineer

Obviously the need for service and maintenance on existing plant exceeds the demand for new installations and for this reason it is essential that all engineers are conversant with the various service operations and diagnostic procedures; they are dealt with in Part One.

The engineer must be conversant with the type and location of system service valves, with the gauge manifold and all types of gauges used with refrigeration equipment.

The ability to diagnose why a system is not operating correctly, or merely to establish that it is providing optimum service, starts with the fitting of gauges to record either the operating or static pressure of the plant, depending upon the circumstances.

A high percentage of service calls are due to refrigerant leakage either from fractured pipework joints or from defective components which have been subjected to internal or external corrosion to cause a leak.

Often a service call can result in the replacement or relocation of a component, necessitating a change in the pipework design or route – so the two aspects, servicing and installation, cannot be completely divorced from each other.

The response to a service call is a venture into the unknown and the engineer must be prepared for every eventuality.

Installation procedures

Part Two starts with the most commonly encountered pipework fittings, methods of joining and supporting pipework and the important subject of oil traps to ensure oil return to the compressor for adequate lubrication at all times.

Installations vary according to system design, system application and the location of major components, of which the most important aspects are discussed. The requirement for ancillary controls and components also depends upon the system design.

The installation of compressors and drive couplings, including the correct alignment of drive options and drive belt tensioning, is of utmost importance, whether in a new installation, in replacement or during commissioning. The necessity to provide electrical protection for both equipment and personnel must not be overlooked.

System evacuation for newly installed systems and after remedial service has taken place must be given priority if reliability is to be achieved. Decontamination of systems often presents problems unless the prescribed and proved procedures are adopted.

For all of these requirements the correct, logical and safe practices are explained. Unless good refrigeration practices are observed, not only does a system's reliability become suspect; there is also the danger of polluting the atmosphere in general, including the immediate vicinity where other people may be affected to the detriment of their health and working conditions.

Vapour compression systems

The basic refrigeration system comprises four major components (Figure 1):

1   The evaporator or cooling coil, which creates a cool surface to which heat may transfer from the refrigerated space or product and be absorbed by the cooling agent (refrigerant) circulated within.
2   The compressor, which circulates the cooling agent and changes its state by compression, creating a pressure differential at B.

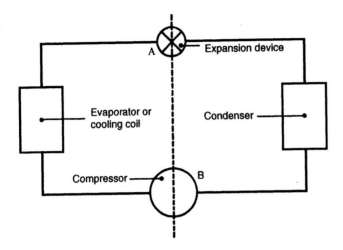

Figure 1 *Basic vapour compression system*

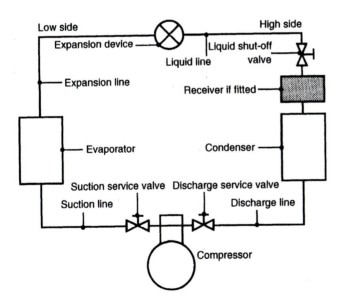

Figure 2 *Vapour compression system*

3   The condenser, which provides a sufficient surface area to be able to reject the heat absorbed by the cooling agent or refrigerant, conveyed from the evaporator.

4   The expansion device or refrigerant metering control, which regulates the flow of refrigerant to the evaporator and creates a pressure differential at A.

Some systems include a liquid receiver, which acts as a storage container for surplus refrigerant during normal operation and is able to retain the bulk of the system operating charge whilst certain service operations are performed.

Refrigerant is conveyed to and from these components by interconnecting pipework, as in Figure 2.

It will be seen from Figures 1 and 2 that the system is divided. The low side runs from the outlet of the expansion device to the inlet of the compressor, and the high side from the outlet or discharge of the compressor to the inlet of the expansion device. The compressor and expansion device are the two components which create a pressure differential within the system, and malfunction of either will affect the system operation.

Obviously, the presence of high and low pressures within the system necessitates the use of valves to isolate those pressures during service operations (Figure 2). These valves are known as service valves and, although the design and position of the valves may vary, their function remains the same.

Part One
Servicing

Since this handbook was first published changes have taken place which affect some of the previous recognized practices.

New and replacement refrigerants and lubricating oils have been introduced and no doubt many more are contemplated. A selection of these are dealt with in this edition.

It will also be noted that some refrigerants which have been phased out and some which will be phased out by the year 2000 are still mentioned. They have been included in this edition for reference and comparison.

Some calculations based on the use of these substances remain to describe the principle. Systems charged with these refrigerants are still operating nationwide and need to be considered.

2 Service valves and gauges

System service valves

When servicing or commissioning any equipment it is necessary to record the system pressures by fitting pressure gauges. To make this possible, all commercial systems generally include at least three service valves: suction, discharge and liquid shut-off.

Suction and discharge service valves may be located on the compressor body of both reciprocating open-type compressors and semi-hermetic motor compressors, but on some compressor designs the service valves may be an integral part of the compressor head assembly. Hermetic motor compressors and some semi-hermetic models do not feature a discharge service valve, but the high side pressure may be obtained from the service valve on the liquid receiver or via a Schraeder-type valve fitted into the discharge line or on the receiver itself.

Service valves can be set to three different positions (Figure 3):

Front seated position   The valve stem is turned fully clockwise to effectively stop the flow of refrigerant vapour from the suction line union on the low pressure side of the compressor and to the discharge line union on the high pressure side of the compressor.

Back seated position   The valve stem is turned fully counter-clockwise to stop the flow of refrigerant vapour to the gauge port of the service valve.

Midway position   The valve stem is turned either clockwise or counter-clockwise to leave the valve unseated. Thus refrigerant vapour can flow from the suction line and also to the discharge line and at the same time pass through the gauge port, to the gauge hose and to the relevant pressure gauge.

The liquid shut-off valve may be located at the outlet of the receiver. It is a single seating valve, i.e. it is either open or closed (Figure 4). On systems which do not have a receiver the valve design will be similar to that of a suction or discharge service valve and will have a gauge connection. When closed or fully front seated, this valve will stop the flow of liquid refrigerant from the condenser or receiver to the expansion device.

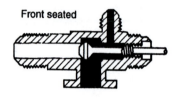

Front seated

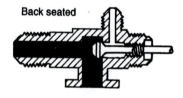

Back seated

Midway

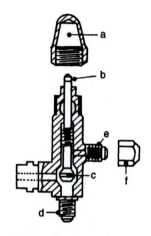

a Valve seal cap
b Valve stem
c Valve stem boss
d Suction or discharge line union
e Valve gauge port union
f Seal cap

Figure 3   *Service valve*

Performing service operations, carrying out repairs, commissioning a system and diagnosing faults involve the use of these valves in addition to test equipment, which will be dealt with separately.

Service gauge manifold

This is also called a system analyser. It consists of a manifold on the top of which are mounted two pressure gauges (Figure 5). Underneath the manifold

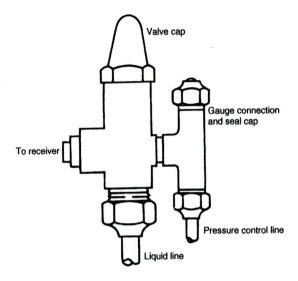

Figure 4 *Typical receiver valve*

are three external hose connections, and at either side of the manifold is a shut-off valve. These valves seat to a position at each side of the centre hose connection, and when closed (turned fully clockwise) will prevent passage of vapour to the centre hose. Current manifold designs may include other features, but in this instance the standard model is illustrated.

A colour code has been introduced for the pressure gauges, hoses and shut-off valves.

The left hand pressure gauge is known as a compound gauge because it will record both positive and negative pressures, since the gauge is calibrated to read zero at atmospheric pressure. The gauge pressure range is from 30 in Hg to 0 psi (0.9 to 0 bar) for pressures below atmospheric pressure, and from 0 to 250 psi (0 to 10.7 bar) or more for those above. This gauge is coded blue.

The right hand gauge is called a pressure gauge. It only records pressures above atmospheric, from 0 to 500 psi (0 to 35 bar). This gauge is coded red.

The hoses are also colour coded to correspond to the gauges and shut-off valves. The blue hose should be connected to the compound gauge, the yellow to the centre connection and the red to the pressure gauge.

When the manifold is assembled it is not necessary to open the valves. Pressure will be recorded as soon as the system pressure is passed to the hose after setting the compressor service valve. When either valve is opened, and

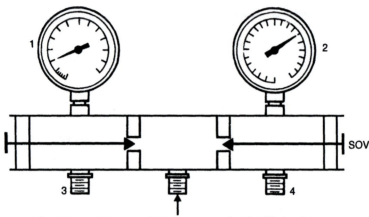

Connect to refrigerant cylinder for vapour charging/discharging
or evacuating system; keep capped if not required

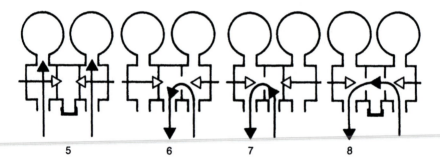

1 compound gauge	5 recording pressures	SSV suction service valve
2 pressure gauge	6 removing refrigerant	DSV discharge service valve
3 hose to SSV	7 adding refrigerant	SOV shut-off valve
4 hose to DSV	8 by-passing	

Figure 5   *Service gauge manifold*

assuming pressure is available from the system, it will pass to the centre hose.
It is advisable to keep the centre hose plugged at all times, or the centre
connection capped when the hose is removed.

With the centre connection capped and both valves open, it will be seen
that the pressures will equalize on both gauges. When both valves are closed
and the centre connection is capped, it will be seen that both negative and
positive pressures can be recorded when the compressor service valves are set

to operating positions, assuming that the system is designed to operate in such a manner.

The function of the valves on the manifold is shown in Figure 5.

The service gauges are a vital part of a service engineer's equipment. They are invaluable for performing service operations or diagnosing faults, which will be dealt with later.

Fitting gauges

Although the task of fitting gauges is a simple one, it must be realized that any refrigerant has a pressure/temperature relationship: the higher the temperature of the refrigerant, the greater the pressure. It will be seen from the pressure/temperature chart at the end of this manual that any refrigerant, if subjected to even a normal ambient temperature, will generate sufficient pressure to present a danger. Thus the engineer should be conversant with the safety procedures for handling refrigerants, and should observe them.

Open-type system

Service gauges may be fitted when the plant is at rest or when it is operating. In the latter case the high side system pressure will be much greater, and it is recommended that the following procedure be adopted for either condition (Figure 6):

1 Remove the valve caps from the suction and discharge service valves on the compressor.
2 Set both valves to the back seat position.
3 Remove the seal caps from the gauge connections of both service valves (normally a flare nut and bonnet).
4 Fit the blue hose from the manifold to the suction service valve connection.
5 Fit the red hose from the manifold to the discharge service valve connection.
6 Ensure that the centre hose connection on the manifold is capped or that the yellow hose is plugged.

Purging hoses and manifold

Having fitted the gauges to the service valves, the next step is to remove any air from the hoses and manifold. Air contains moisture and is a contaminant, and its entry to the system must be prevented.

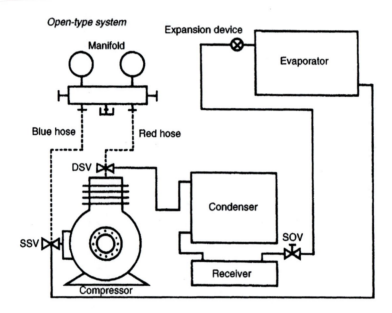

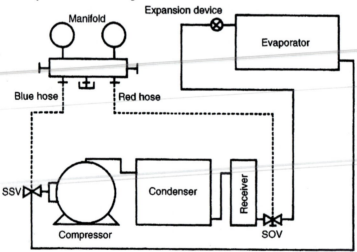

SSV suction service valve
DSV discharge service valve
SOV shut-off valve

Figure 6   *Gauge manifold connection*

Once the gauges have been fitted, proceed as follows: .

1   Crack off the discharge service valve (turn the valve stem one turn clockwise).
2   Open both shut-off valves on the manifold.
3   Loosen the hose connection (blue) on the suction service valve and allow the pressure to leak out slowly for one or two seconds.
4   Tighten the hose connection.
5   Close both shut-off valves on the manifold.
6   Crack off the suction service valve.

Further points on fitting

When fitting gauges to service valve gauge connections, always use two spanners of the correct size to avoid breaking the vapour seal of the gauge unions.

The plant, if operating with a low pressure control, may have the control tubing or capillary connected with a tee union to which the gauge hose is also fitted. Thus when the service valve is back seated, the pressure will be trapped in the control line. When pressure is released during the purging operation, the plant will stop; it will restart when the suction service valve is cracked off to pressurize the control.

When gauges are fitted to a plant which has been closed down, the liquid shut-off valve must be opened so that refrigerant will circulate around the system.

System without a discharge service valve

The procedure for fitting gauges is the same, except that the red hose will be connected to the gauge connection of the service valve on the receiver (Figure 6). When this valve is back seated, the refrigerant will circulate.

Recording

After fitting gauges allow a short time to elapse before recording the pressures, i.e. when the pressures stabilize on the gauges.

Removing gauges

Once again, bearing in mind that high pressures are generated whilst a plant is operating, care must be taken not to pollute the atmosphere when gauges are removed. The following procedure will minimize the loss of refrigerant and discharge to atmosphere:

1 Back seat the discharge service valve.
2 Open both shut-off valves on the manifold and allow the pressures to equalize in the gauges.
3 Close high side valve on the manifold and operate the compressor until a zero pressure is indicated on the compound gauge. Close the low side valve on the manifold.
4 Remove the hose from the discharge service valve union, replace the seal cap and tighten.
5 Back seat the suction service valve.
6 Remove the hose from the suction service valve union, replace the seal cap and tighten.
7 Set both service valves to the operating positions (cracked off the back seat).
8 Replace the service valve caps and tighten.
9 Test for leaks at the valve caps and gauge unions (see Chapter 3).

The low side pressure in systems charged with refrigerants R22 and R502 can be much higher than that in systems charged with refrigerant R12. Thus when removing the hose from the low side, loosen the connection and release the pressure slowly. Although the pressure may be high, the volume of vapour in the hose will be small.

Pressure controls

It is common practice for systems to be installed with pressure safety controls. When gauges are removed, the service valves must be left in the operating positions – cracked off the back seat – to allow the controls to be pressurized (see Chapter 6).

Other valves

Schraeder valves

Modern service manifolds incorporate additional hand valves, connections for a vacuum pump, and a liquid-indicating sight glass. Some feature Schraeder

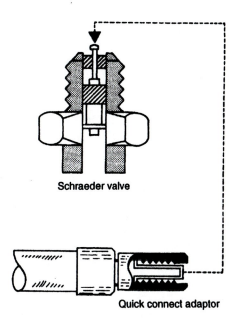

Schraeder valve

Quick connect adaptor

Figure 7 *Schraeder valve*

valves in addition to the normal hand valves on the manifold. These additional valves make it possible to perform various operations without the necessity for removal of the manifold or hoses.

The Schraeder valve is similar to that used in cycle or car tyres (Figure 7). The fitting into a system can vary: it may be brazed in, screwed into a component or supplied as a flare connector.

When connection is made to this type of valve it is necessary for the service hose to include a quick release adaptor. This depresses the valve stem and allows passage of refrigerant into the hose.

Line tap valves

These are sometimes referred to as piercing valves. They consist of a small clamp which is in two sections fitted to the pipework.

Totally sealed systems will involve the use of this type of valve in order to gain access to the system. A special charging/gauge adaptor is screwed on to the piercing valve assembly (see Figure 8).

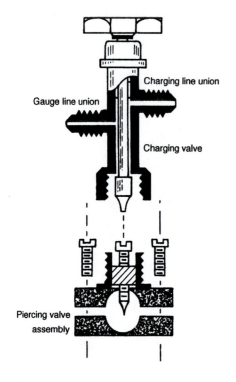

Figure 8   *Line tap valve*

The valve assemblies are available for a wide range of pipe sizes and the adaptor is a standard fitting to the assemblies.

3 *Refrigerant leak detection*

A leak test is a vital operation after the installation or service of refrigeration equipment. An efficient leak test will prevent expensive call-backs and losses.

Shortage of refrigerant in a system results in reduced plant capacity and thus in longer running times of the equipment, which could prove costly to the customer over a prolonged period. It can also be the cause of a major breakdown of a compressor due to inadequate lubrication and cooling, as explained in Chapter 10.

A system must be operating with a full charge of refrigerant to enable a commissioning engineer to obtain the design conditions by the setting of various controls (see Chapter 16).

Dye systems

Some plants may be charged at the time of installation with a refrigerant which contains a coloured dye to provide a visual indication of leakage. Systems using refrigerant containing a dye must always be liquid charged. This is most important when adding refrigerant to a system after a considerable loss due to leakage.

The dye tends to separate from the refrigerant in the cylinder and is heavier than the liquid refrigerant. Therefore it is advisable to agitate the liquid when charging or adding refrigerant by placing the cylinder in a horizontal position and rocking it gently.

Refrigerant charging procedures are dealt with in Chapter 4.

Test pressures

During normal service operations a leak test does not entail a prior evacuation of the system unless the system is contaminated or the refrigerant has been completely discharged. It is essential however, that a minimum of 30 psig (pounds per square inch gauge) or 2 bar exists in the system when testing for leaks. If no leaks are found then the system should be tested again at operating pressures.

A pressure leak test is more satisfactory and will be dealt with later.

Methods of testing

Note: Refer to Chapter 16 on refrigerants and revised procedures for leak detection, evacuation during commissioning.

Bubble test

The most common and inexpensive test is the bubble test. A water/soap solution is simply brushed around a joint or component which is suspected of leaking, or sprayed on with an aerosol. It is recommended that proprietary rather than made-up solutions are used, as they are more viscous and the bubbles are stronger and longer lasting. Ordinary soap bubbles are weaker and are normally short lived.

The disadvantage of this method is that a large leak can blow through the solution and then no bubbles will appear, although in most such cases the leak will be audible.

Halide torch

This consists of a small burner assembly mounted on top of a container of gas, for example propane. The burner comprises a hand valve and a venturi or mixing chamber with an attachment for the exploring tube. Above the orifice of the burner there is a copper ring, a strip or a tube through which the flame passes when the torch is ignited.

When the torch is lit the air will be drawn into the venturi via the exploring tube, and the flame will burn slightly blue or colourless. When a trace of halogen refrigerant (R11, R12, R22, R500, R502 etc.) mixes with the air, the flame will immediately change colour as the refrigerant vapour contacts the hot copper ring or tube. The colour will range from green for a small leak to dark blue or purple for a large leak. When the refrigerant burns off, a toxic atmosphere will be created.

To test for leaks the exploring tube is passed around the suspected area slowly, and for effectiveness the plant should be stopped.

Electronic leak detector

This is the most sensitive type of leak detector, and many designs are available. Some respond to an ion source, and others to a change of temperature (thermistor); the dielectric type is based on the conductivity of different gases.

These instruments are dry cell battery operated. When used the sensor or sensing tip should be inspected for cleanliness; tips should always be kept free from dirt and lint. Filters should be changed regularly, because a contaminated filter will cause the instrument to respond as if a leak was detected.

Normally the instruments will respond to atmospheric air by giving out an audio signal (bleep) at approximately one bleep per second. When the halogenated refrigerant contacts the sensor the signal will accelerate, depending upon the degree of vapour leaking; a large leak could produce a continuous signal or oscillation.

One disadvantage is that because of the sensitivity of the instrument it will respond to minute volumes of refrigerant vapour, and sometimes it will prevent the actual pin-pointing of a leak. It is also responsive to expanded foam insulants, thereby making somewhat difficult the detection of leaks on pipework passing through coldroom walls and parts of domestic appliance systems.

When using a detector the amount of air movement must be reduced to a minimum, i.e. all fans should be switched off and draughts excluded. The sensing tip should be applied below a joint because refrigerant vapour is heavier than air, and then moved slowly around the area.

Nessler's reagent

This is a chemical solution used to detect leaks on water cooled systems charged with ammonia, which has an affinity for water. The solution is added to the recirculated water. If ammonia is present in the water, the solution will react to the nitrates contained therein to change the colour of the water to brown.

Sulphur candle

This takes the form of a small candle or taper which, when ignited, will smoulder and give off sulphur fumes. It is also used to detect leaks on pipework and components in ammonia systems. When it is passed around a joint suspected of leaking, the ammonia fumes will mix with the sulphur fumes to produce a white vapour.

Both ammonia and sulphur fumes are toxic. Thus adequate ventilation must be provided, and precautions must be taken not to inhale the fumes. Ammonia leaks are easily detected by its pronounced odour at 3 to 5 parts per million. At approximately 15 ppm the vapour is very toxic, and at 30 ppm a suitable respirator will be required. Ammonia is lethal at 5000 ppm, and the maximum exposure time to atmospheres of 50 ppm is 5 minutes.

Ammonia becomes flammable at 150 000 to 270 000 ppm.

Halogenated refrigerants

These are compounds which contain one or more of the halogens, such as fluorine, chlorine, iodine and bromine. When they are exposed to a naked flame they will burn and produce a pungent odour, which can be injurious to the human respiratory system.

Pressure leak testing

This is carried out on new system installations or when a plant has been discharged of refrigerant prior to repair. It involves the use of oxygen-free nitrogen (OFN) which is a high pressure gas. This is used to obtain a higher pressure than that of the refrigerant in normal ambient temperatures. This pressure should be controlled and in excess of that under which the system is expected to operate with a normal operating charge of refrigerant. This could be as high as 500 psig (33 bar) in some instances.

Sometimes it is the practice to test the system pipework only in this manner before connecting to the condensing unit. If the condensing unit is new it would have already undergone very stringent tests to pressure vessel standards by the manufacturer.

When an installation is completed and a pressure test is to be carried out it is most important to ensure that the compressor is isolated, irrespective of design, before pressurising the system. The suction service valve should be front seated to avoid any damage to the compressor valves. This will also prevent rupture of the crankshaft seals in open type and semi-hermetic motor compressors.

All pressure controls **must be disconnected or by-passed**. If the expansion valve is not capable of withstanding the test pressure then it too must be removed or by-passed. The bellows or diaphragm of an expansion valve has a maximum operating pressure which must not be exceeded.

The OFN cylinder **must be fitted with an approved regulator** to control the test pressure. For safety reasons a pressure relief valve preset to the test pressure is recommended.

When the system is pressurized, that pressure should be recorded and the plant left for a reasonable period. Just how long that period should be is debatable. It could take a considerable time for a noticeable drop in the nitrogen pressure to become evident when a system has a small leak. The time allocated by some installers and service outlets varies, and with large installations periods of days under pressure is not uncommon.

With the time element being an important factor a 'bubble test' may be permissible but this is not accepted by all suppliers of refrigerants.

British Standard 4434 1995 states that **only an inert gas may be used for pressure testing**. R22 or any other refrigerant is not to be used as a 'trace' pressure test. The 'trace' is no longer regarded as good practice. (For reference, the trace method of leak detection involved charging a system with a small amount of its operating refrigerant and boosting the pressure within the system with nitrogen. The leak could then be detected using a halide torch. Another method of leak detection is to draw a vacuum on the complete system but again the waiting period will be necessary to see if the vacuum is held. The disadvantage of drawing a vaccum would be the ingress of air which contains moisture if a slight leak were to break the vacuum.

Handling refrigerants: safety precautions

Although the common refrigerants (R12, R22, R502 etc.) are not considered hazardous, it must be remembered that all refrigerants are heavier than air and will replace air in a confined space very quickly. This can be dangerous; if the air does not contain at least 19 per cent oxygen, loss of consciousness may result.

When testing for leaks, always ensure that the area is well ventilated if at all possible. Always stand to one side of the detector in case there is a sudden violent discharge from the suspected pipework.

The following precautions should be taken:

1 Wear goggles, gloves and overalls at all times to protect eyes and to prevent direct contact of refrigerant with the skin, which can cause burns. This applies especially when charging or discharging refrigerant.
2 Make sure that the service cylinder is not overfilled.

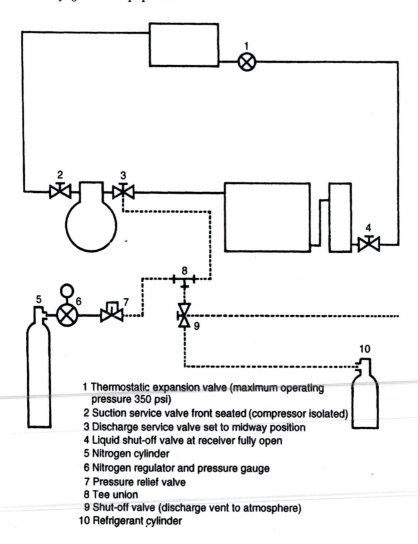

1 Thermostatic expansion valve (maximum operating
   pressure 350 psi)
2 Suction service valve front seated (compressor isolated)
3 Discharge service valve set to midway position
4 Liquid shut-off valve at receiver fully open
5 Nitrogen cylinder
6 Nitrogen regulator and pressure gauge
7 Pressure relief valve
8 Tee union
9 Shut-off valve (discharge vent to atmosphere)
10 Refrigerant cylinder

Figure 9   *Typical test arrangement for leak testing at 300 psi*

3   Do not expose cylinders to direct sunlight, radiated heat or convected heat
   from appliances.
4   Avoid discharge near naked flames or flame producing appliances.
5   Avoid direct contact with refrigerant/oil solutions from hermetic systems
   (motor burn-out), which can be very acidic.

6 Always vapour charge a system from the low side to avoid possible damage to compressor valves.
7 Always check that the refrigerant is correct for the system being charged.
8 Whenever possible ensure that the working area is well ventilated. If ammonia is being used, ensure that a respirator or some form of breathing apparatus is at hand.

Work safely!

4 Pumping down and charging

Before an engineer can repair a refrigerant leak in the system it has to be located and for this reason leak detection is the first step. If the system has not lost all its refrigerant charge, leak detection may proceed as described in Chapter 3 provided that there is an adequate test pressure within the system. When the entire refrigerant charge has been lost, the system has to be pressurized.

If the leak is found to be minor and on the low side of the system, 'pumping down' will be necessary before any attempt is made to repair the leak. Leakage from the high side of the system will require the removal of all the refrigerant before repairs are undertaken.

Pumping down

This procedure transfers the refrigerant circulating around the system from the liquid shut-off valve at the liquid receiver to the inlet of the compressor for storage in the condenser and the receiver, thereby making it possible for other tasks to be performed.

The method is as follows:

1 Fit gauges, set the service valves to operating positions and operate the plant.
2 Close the liquid shut-off valve at the receiver.
3 Allow the compressor to operate until a pressure slightly above atmospheric pressure is registered on the compound gauge (3 psi or 0.2 bar). If the system employs a low pressure cut-out switch then the range will need to be altered; note the cut-out pressure beforehand. Modern switches have a small lever which can be moved to override the cut-out point, thus keeping the compressor running.
4 Stop the plant and front seat the suction service valve.

The system can now be worked on for a number of repairs.

When it is required to close down a plant for an extended period, the discharge service valve should also be front seated after the gauges have been removed.

Adding refrigerant

When it is necessary to add refrigerant to a system after a loss it indicates that there is a leak in the system. The leak must be located, repaired and the system leak tested. If the leak is found to be on the low side of the system the repair can be made after the system has been 'pumped down' to a balance in pressure. **Do not draw in any air which contains moisture, especially where hygroscopic ester oils are concerned, (for example R134a)**. Some oils are not affected and accept the ingress of air and moisture, recovering after a vacuum is drawn with a recovery unit.

Vapour charging

A slight loss of refrigerant is recognized by 'bubbles' in the liquid-indicating sight glass installed in the liquid line of a system, a loss of performance, partial frosting of the evaporator and lower than normal operating pressures. Refrigerant may be added in vapour form to the low side of the system if the refrigerant is azeotropic such as R502. This means that it reacts as a single substance refrigerant.

The zeotropic refrigerant blends do not react in a similar manner and they should be added to a system in liquid form. This subject is covered in more detail in Chapter 17.

The procedure for vapour charging is as follows (Figure 10):

1   Fit gauges, set the service valves to operating positions and operate the plant.
2   Obtain a service cylinder of the correct refrigerant: this can be verified from the equipment log, the compressor nameplate or the label on the expansion valve.
3   Connect the yellow hose to the centre connection on the manifold and to the service cylinder.
4   Open the valve on the service cylinder, loosen the connection on the centre hose on the manifold and purge air from the hose.
5   Tighten the hose connection and set the suction service valve to the midway position.

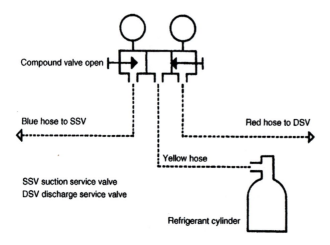

Compound valve open

Blue hose to SSV

Red hose to DSV

Yellow hose

SSV suction service valve
DSV discharge service valve

Refrigerant cylinder

Figure 10 *Vapour charging*

6 Open the compound gauge valve on the manifold slowly, and regulate the refrigerant into the system at an approximately average suction pressure (e.g. 30 psi or 2 bar for R12).

7 Observe the liquid-indicating sight glass and, when the bubbles cease, close the compound gauge valve on the manifold. If bubbles return intermittently after a short time, add more refrigerant. When bubbles have ceased completely the operating pressures will have returned to normal and the evaporator will be fully frosted.

A further check on the refrigerant charge is described on pages 33–8.

When vapour charging a system with an azeotropic refrigerant such as R502 the refrigerant cylinder must always be kept in a vertical position to prevent the possibility of liquid refrigerant from entering the compressor. This can create a dynamic pressure when the compressor starts, causing damage to valves or may even break piston connecting rods and damage pistons. The liquid refrigerant will also flush lubricating oil from bearing surfaces.

The main disadvantage of vapour charging to the low side of the system is that it is a comparatively slow process, especially during low ambient temperature conditions and when the system requires a large operating charge. A large compressor will quickly reduce the suction pressure. The cylinder will become cold and, if the low pressure condition is prolonged, the cylinder will frost up.

Hermetic and semi-hermetic motor compressors may be suction vapour cooled; that is, they may rely upon suction vapour returning to the compressor in sufficient volume to cool the motor windings. It is possible that the low volume of vapour entering the compressor during the charging process will not be adequate for motor cooling.

The charging process can be speeded up by applying swabs dipped in warm water to the refrigerant cylinder. Never heat cylinders with a blowtorch or immerse them in boiling or very hot water.

A rapid charger is available. This device, fitted between the hose and the cylinder, acts as a restrictor or limit valve, maintaining a higher constant pressure in the cylinder and preventing early frosting. The vapour passing through the device is less superheated when it reaches the compressor, so less time is taken to charge the system.

Liquid charging

This method is not always acceptable to manufacturers, since it involves putting refrigerant liquid into the high side of the system. If the compressor discharge valves are not seating properly there is a danger of liquid refrigerant entering the compressor cylinders, which can cause damage due to dynamic pressure when the compressor starts. The liquid cannot be compressed without creating high pressure.

Some large condensing units are equipped with liquid charging valves on the receiver. Smaller hermetic and semi-hermetic units have service valves located on the receivers, and these can be employed for charging.

Liquid charging is carried out when a system is commissioned, or when it has been completely discharged of refrigerant.

Compressors should be at rest when the system is being liquid charged.

A simple method is by gravity, and should be performed carefully for the initial liquid charge (Figure 11a):

1 Fit gauges and connect to a refrigerant cylinder as shown.
2 Fully open the charging valve or service valve.
3 Open the valve on the refrigerant service cylinder and invert the cylinder. As long as the pressure in the service cylinder is greater than that in the system, the liquid will flow. The flow will be audible.
4 When the flow has ceased, allow gauge pressures to stabilize, close the valve on the service cylinder and operate the plant.
5 Complete the charging in vapour form to the low side of the system.

To eliminate the possibility of compressor damage, the following procedure may be adopted (Figure 11b):

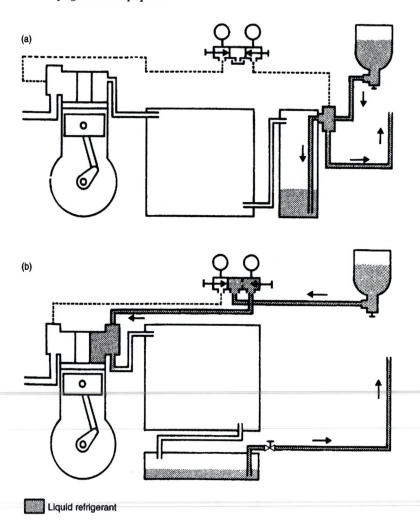

Figure 11   *Liquid charging*

1 Run the compressor for a few revolutions with the discharge service valve
   fully front seated until a pressure of 20 to 30 psi (1.5 to 2.0 bar) is registered
   in the pressure gauge. This will be compressor cylinder pressure.
2 Stop the compressor.
3 Invert the refrigerant cylinder and open the valve or open the liquid
   charging valve as appropriate.

Table 1 *Initial charges (kg) for various compressors and refrigerants and for flooded and dry evaporators*

Compressor capacity kW	R12		R22		R502	
	Flooded	Dry	Flooded	Dry	Flooded	Dry
0.5	1.4	0.7	1.4	0.7	1.4	0.7
0.75	2.7	1.4	2.7	1.4	2.7	1.4
1.0	4.0	2.0	4.0	2.0	4.0	2.0
1.5	5.5	2.7	5.5	2.7	5.5	2.7

4 Open the liquid charging valve or service valve on the receiver.
5 Crack off the discharge service valve. High pressure from the compressor on the surface of the liquid refrigerant in the cylinder will force the liquid into the condenser and receiver.
6 Set the discharge service valve to the operating position, close the liquid charging/service valve and the refrigerant cylinder valve, and operate the plant.
7 Complete the charge in vapour form.

With water cooled condensers the pressure will normally be greater in the refrigerant cylinder than in the system.

As a guide to refrigerant charges, Table 1 gives approximations for the initial charges to be administered to systems employing both flooded and dry expansion evaporators.

Liquid charging is to be carried out more frequently since the new and replacement refrigerants become available. The current and more efficient method is when the refrigerant is added via the liquid receiver as shown in Figure 11b.

Refrigerant cylinders and fill ratio

Refrigerant bulk cylinders and some service cylinders are fitted with two shut-off valves, which are colour coded and marked to indicate delivery of either liquid or vapour.

Two tubes supply refrigerant to the valves (Figure 12). One terminates at the top of the cylinder above the liquid level, and when the vapour valve (red) is opened it will discharge vapour only. The other tube from the liquid valve

Red (vapour)   Blue (liquid)

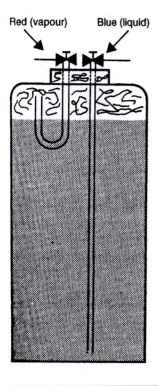

Figure 12 *Refrigerant cylinder*

(blue) terminates almost at the bottom of the cylinder, and when the valve is opened it will discharge liquid until the cylinder is virtually empty.

Modern cylinders have capacities, maximum fill quantity and identification of the contents stamped on the cylinder. Older types may only have the total weight and the tare (empty cylinder weight). The total weight is the weight of water contained in the cylinder if it was completely full (the water capacity) plus the tare.

A refrigerant is a volatile substance, and is more responsive than water to density changes at different temperatures. For this reason a cylinder should never be completely filled with refrigerant.

A simple calculation can be made to determine the amount of refrigerant which can safely be transferred into a service cylinder. A safe fill for any refrigerant is 80 per cent of the water capacity. For example, assume the total weight of a cylinder to be 24 kg and the tare to be 4 kg. Then the water capacity is 24 − 4 = 20 kg. The maximum fill is 80 per cent of 20 kg, i.e. 16 kg.

5 Operating principles and symptoms

When a system has developed a refrigerant leak, the repair has been effected, it has then undergone a final leak test and has been charged with refrigerant, the efficient service engineer will always check the operating conditions of the plant. This chapter is designed to assist the engineer in establishing whether the plant is indeed operating to its design conditions.

Merely to repair a leak and recharge the system may be a 'short cut to disaster' if the refrigerant leak was in fact the result of another, undiscovered fault which will lead to an expensive call-back and a dissatisfied customer: short cuts can lead to trouble!

Service gauges play an important role in diagnosing faults on refrigerating equipment. Low suction pressure can be related to a number of faults. It is normally associated with restricted refrigerant flow to the evaporator, but in some cases it can be due to the improper setting of a temperature control.

As well as maintaining a design temperature, it is vital that the correct humidity conditions are achieved for the type of product being stored.

Temperature, humidity and air motion in food storage

It is important that the temperature difference (TD) between the product being stored and the refrigerant is correct for that product. Too wide a TD will result in excessive dehydration of the product. Too close a TD can result in rapid deterioration of the product; fresh meat, for example, will soon become discoloured and slimy to the touch. This section deals with storage conditions and temperature control settings, thereby giving an indication of evaporating temperatures and pressures to be expected during system operation.

Food products may be divided into four classes to provide proper storage conditions:

1   Foods which dehydrate quickly: fruits, vegetables, eggs and cheese.

Table 2 *Recommended TDs for four food classes and two evaporator types*

Food	Gravity coil		Forced air coil	
	°F	°C	°F	°C
1	15 to 20	8.5 to 11	6 to 8	3.3 to 4.4
2	18 to 28	10 to 15.6	10 to 12	5.6 to 6.7
3	20 to 25	6.0 to 13.9	12 to 15	6.0 to 8.3
4	27 to 37	15 to 20	15 to 25	8.3 to 13.9

2 Foods subject to sweating and some dehydration: fresh cut meats and provisions.
3 Carcase meats, chilled meats, and products not subject to excessive dehydration.
4 Products not subject to dehydration: dried fruits, tinned goods and canned/bottled beverages.

Table 2 gives the recommended TDs for these classes of product. Slight variations may be necessary to obtain ideal conditions.

For better understanding, a few examples of the simple calculations needed to determine various factors are now given. These calculations are based on a system charged with refrigerant R12. See Figure 13.

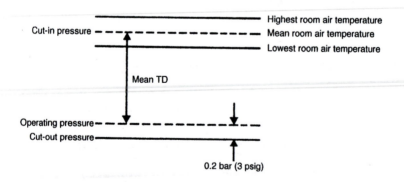

Figure 13 *Temperature difference*

To determine the TD

1   Obtain the suction pressure at which the unit cuts out, add 3 psig or 0.2 bar to this pressure and convert to temperature. This will be the average evaporating temperature.
2   Subtract the average evaporating temperature from the product temperature. This will be the TD at which the system is operating.
3   Compare this with that recommended for the product classification.

Example

For chilled meat (product class 3), at a storage temperature of 32 to 36 °F (0 to 2.2 °C) and using a forced air evaporator, the recommended TD is 12 to 15 °F (6.0 to 8.3 °C). The average product temperature is therefore 34 °F (1.1 °C).

The cut-out pressure is 22 psig (1.5 bar), and so the average suction pressure is $22 + 3 = 25$ psig $(1.5 + 0.2 = 1.7$ bar$)$. This converts to an average suction temperature of 22 °F or −5 °C. The TD is therefore $34 − 22 = 12$ °F $(1.1 − (−5) = 6.1$ °C$)$. It can be seen that the TD is within the recommended range.

To establish the low pressure cut-out point

1   Subtract the TD from the product temperature to obtain the average evaporating temperature.
2   Convert this temperature to pressure, which will be the average evaporating pressure.
3   The cut-out pressure will then be 3 psig or 0.2 bar below the average evaporating pressure.

Example (imperial)

For a class 3 product with a storage temperature of 30 °F, the recommended TD is 20 to 25 °F for a gravity coil. The average TD is 23 °F. The average evaporating temperature is therefore $30 − 23 = 7$ °F. Converted to pressure, this is 13 psig. The pressure control cut-out point will then be $13 − 3 = 10$ psig.

For an off-cycle defrost with a low pressure control, set the cut-in to 35 psig. This will provide an approximate coil temperature of 38 °F, when the coil will be completely free of frost and ice.

Example (SI)

For a class 3 product with a storage temperature of −2 °C, the recommended TD is 8 °C for a forced air evaporator. Then the average evaporating

Table 3   *TDs between air and evaporator at various relative humidities*

Relative humidity	Gravity coil °C	Forced air coil °C
90 to 95	7 to 9	2 to 4
85 to 95	9 to 13	4 to 6
75 to 85	13 to 15	6 to 9
65 to 75	15 to 17	10 to 12

temperature is $-2 - 8 = -10\,°C$. Converted to pressure, this is 1.2 bar. The cut-out pressure is established as $1.2 - 0.2 = 1$ bar.

When selecting cut-in and cut-out pressures to control fixture temperatures, the following information is required:

1 Dry bulb temperature of refrigerated space.
2 Relative humidity of refrigerated space.
3 Type of refrigerant used.
4 Type of evaporator.

Table 3 shows TDs between air and evaporator for various humidities.

Calculating the operating head pressure

Manufacturers of refrigeration condensing units will supply technical data for their products on demand, but this will be based on ideal operating conditions in controlled ambient temperatures. The service engineer or technician, dealing with many different types of equipment, needs a simple quick method to enable him to determine the theoretical operating head pressure of a plant so that this can be compared with the actual operating pressure to assist in fault diagnosis.

It is known that a shortage of refrigerant will produce a low suction pressure, and the fact that the compressor is doing very little work in handling vapour at a lower density will affect the pressure ratio of the compressor.

A restricted supply of refrigerant to the evaporator will not affect the operating head pressure so dramatically as an acute shortage of refrigerant, because there will be a higher pressure existing in the high side of the system, which now contains almost all of the refrigerant charge in the condenser and the receiver. It is therefore possible that only a slightly lower pressure than that to be expected will be registered in the pressure gauge.

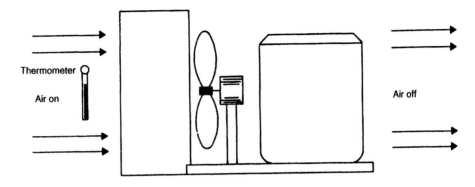

Figure 14 *Taking the temperature of the condensing medium*

The main factor which determines the operating head pressure of a plant is the temperature of the condensing medium, which is the air passing over an air cooled condenser or the water in a water cooled system. It is important that the temperature of the condensing medium be taken accurately and from the correct location. For an air cooled unit this will be the air on to the condenser and not the air off, because the condenser will have rejected heat to the air passing over the condenser coils (see Figure 14).

A simple rule of thumb for calculating the operating head pressure of a refrigeration system is as follows:

1 Hold a thermometer in the air stream to the condenser for 2 to 3 minutes and then note the temperature.
2 Add a condensing factor of 15 °C (30 °F) to this temperature and then convert it to pressure by using a refrigerant comparator or a pressure/temperature graph, or by direct conversion from the pressure gauge on the manifold.

This will be an approximate theoretical head pressure, within 0.7 to 1 bar or 15 psig of that found under the conditions the plant is operating.

An excessively high operating pressure is an indication of a condensing problem; insufficient heat is being rejected by the condenser.

When taking the temperature of the water passing through a water cooled condenser, ensure that the temperature is that of the water actually entering the condenser.

Some semi-hermetic motor compressors are fitted with a water coil for compressor cooling. The supply water passes through this coil before it enters the condenser coil or tubes (see Figure 15).

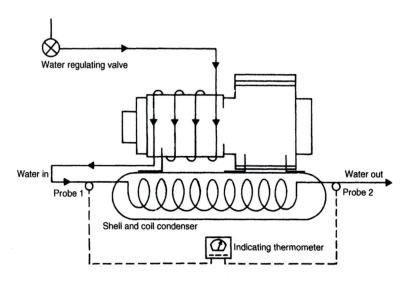

Figure 15 *Water cooling of semi-hermetic motor compressor*

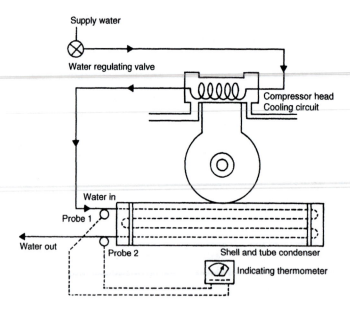

Figure 16 *Water cooling of open compressor*

Some reciprocating open-type compressors used with water cooled systems incorporate a cylinder head cooling feature in the form of a water circuit through the compressor head. Supply water enters this before entering the condenser (see Figure 16).

A rapid indicating thermometer is recommended for taking water temperatures because the water regulating valve responds to the operating head pressure and will vary the temperature of the water as it modulates. It is also necessary, when adjusting the water regulator, to be able to take inlet and outlet temperatures quickly. Locate the thermometer probes tightly to the pipework to ensure good thermal conductivity, note the average inlet temperature and calculate as previously described.

To ensure an economical and adequate water flow through a water cooled condenser, the water regulating valve should be adjusted to provide a temperature difference of 15 to 18 °F or 7 to 9 °C between the inlet and outlet of the condenser.

The thermometer probes should be located as indicated in Figures 15 and 16.

Some examples of head pressure calculations follow. The condensing factors used are 15 °C and 30 °F.

Example

An air cooled condensing unit charged with refrigerant R502 operates with the air on to the condenser at 20 °C. Then 20 + 15 = 35 °C: R502 at 35 °C will register a pressure of 13.8 bar.

If the temperature of the air was lower at 15 °C, then 15 + 15 = 30 °C: R502 at 30 °C will register a pressure of 12.2 bar.

Example

A water cooled condensing unit charged with refrigerant R12 operates with water entering the condenser at 50 °F. Then 50 + 30 = 80 °F: R12 at 80 °F will register a pressure of 98 psig.

If the temperature of the water is higher at 65 °F, then 65 + 30 = 95 °F: R12 at 95 °F will register a pressure of 115 psig.

Compressor pressure ratio

This is the ratio between the suction pressure and the discharge pressure. No refrigeration compressor is 100 per cent efficient, owing to various losses. It is considered that a compressor is 80 to 85 per cent efficient with pressure ratios of 4:1 to 6:1.

With a suction pressure of 2 bar, the operating head pressure could be between 8 and 12 bar depending upon the efficiency of the compressor. If the suction pressure was 25 psig then an expected operating head of between 100 and 150 psig would be expected.

Naturally the choice of refrigerant and the system application must be taken into consideration; the values given here refer to a typical single stage reciprocating compressor.

Symptoms of system faults

When a plant has been charged it must be established that the charge is in fact complete. The disappearance of bubbles in a sight glass does not necessarily mean that the evaporator is correctly flooded.

Continued adding of refrigerant when bubbles show in the sight glass can also result in the system being overcharged.

Shortage of refrigerant in evaporator

A situation could arise where the sight glass shows full of liquid yet a shortage of refrigerant is evident in the evaporator. This may be due to a number of factors.

Refrigerant liquid can flash off in long liquid line runs. The ideal location for a sight glass is just before the expansion valve, although it is common practice to install them close to the condensing unit. Ideally two sight glasses should be installed, one near the condensing unit and the other before the expansion valve. This will determine whether a solid column of liquid reaches the expansion valve.

Restricted refrigerant flow to the evaporator can be caused by the following:

1   A partial blockage may occur in the filter drier, thereby creating a pressure drop in the liquid line. If the sight glass is located after the filter drier, bubbles will be evident. When a sight glass is installed before the filter drier, a pressure drop will exist just the same but bubbles will not be visible. In each case a temperature difference will exist either side of the restriction.
2   An expansion valve filter or screen may be blocked by undesirable substances circulating around the system which have passed through the filter drier, such as moisture or particles of the desiccant in the filter drier. Carbon may build up in the fine mesh of the valve screen.

3 The thermostatic expansion valve may be incorrectly adjusted or defective through partial loss of the phial charge, and therefore will not open sufficiently. A total loss of the valve phial charge will result in a complete blockage and a starved evaporator. Some expansion valves have replaceable cartridges and screens. The cartridge is stamped with an orifice size; too small an orifice can result in a starved evaporator.

4 Some plants employ evaporator defrost systems which incorporate a magnetic valve or solenoid valve. This valve, installed in the liquid line, will stop refrigerant flow to the evaporator when a defrost period is initiated by a timing device. This enables the evaporator to be evacuated of refrigerant so that, when the defrost heaters have completely cleared the evaporator of frost, an excessive build-up of pressure will be prevented during the period of continued application of heat.

Each of these conditions will result in lower than normal operating pressures.

When there is a shortage of refrigerant, pressures will be lower than normal. However, the reduction in pressure may be so slight as not to be readily detected, other than by a loss of evaporating capacity and the longer running time of the unit. If the shortage is considerable, both suction and discharge pressures will be very low; if the system temperature control consists of a thermostat only, the unit will run continuously with poor refrigerating effect. When a low pressure switch is in circuit, the compressor will short cycle (cut in and out quickly) on this control.

Likewise, any restriction in the refrigerant supply will produce the same symptoms on the low side of the system. It is possible for the compressor to operate continuously on a deep vacuum if a low pressure switch is not used. A complete blockage, preventing refrigerant from entering the evaporator, will cause a low pressure switch to stop the compressor; it will not restart because there will be insufficient pressure rise from the evaporator to actuate the switch.

High suction and low discharge pressures

This condition is usually due to a fault within the compressor, such as broken valve reeds or incorrect seating of the valve reeds.

If the suction reeds are at fault, some of the discharged vapour will be forced back into the suction side of the compressor as the piston reaches the top of its stroke.

Faulty discharge reeds will contribute to longer running time, poor refrigeration effect and lower than normal operating head pressure. During an off cycle the discharged vapour will leak back into the cylinder(s), causing

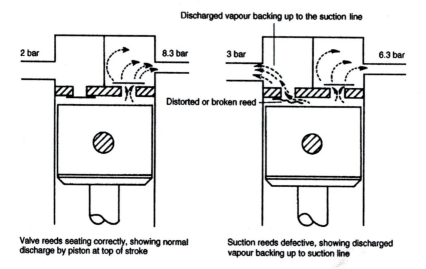

Figure 17   *Valve reeds*

a pressure rise which will actuate a low pressure switch and quickly restart the compressor if this is the temperature control.

Figure 17 shows the passage of refrigerant through compressor valves with serviceable and defective valve reeds.

Compressor valves can become distorted and fractured if liquid refrigerant is allowed to enter the cylinder; this is known as liquid slugging. For this reason service cylinders should never be inverted during the low side charging of a system.

Valve seats and reeds can become pitted if metal filings or small particles of grit are allowed to enter pipework during installation or repairs. They may circulate with the refrigerant and become trapped between the reeds and valve seats when compression takes place.

Excessive wear on pistons and cylinder walls will also contribute to a loss of compression. This is most likely to be experienced on a plant which has been in service for a long period, but could be related to lack of proper lubrication of the compressor.

A shortage of refrigerant can be the cause of oil starvation in the compressor and low pressure in the evaporator. The reduced amount of liquid refrigerant flowing through the evaporator is insufficient to move the oil and return it to the compressor, and it remains entrained in the evaporator.

6 Service diagnosis and repairs

As previously stated, there may be more than one fault in a system and some may persist until such time as a total breakdown or component failure occurs, to the detriment of the product being stored.

For example, after a leak has been discovered and repaired, the system is ready to be charged with refrigerant: reference has already been made to a check of operating pressures and a full sight glass as indications that the system is fully charged. However, this does not necessarily mean that the evaporator is correctly flooded and is operating to its full capacity. To make certain of this, the superheat setting of the expansion valve must be checked. An incorrectly adjusted valve can lead to starvation of liquid refrigerant to the evaporator, or to flooding of the evaporator with the possibility of liquid refrigerant entering the compressor to cause damage.

There is also the possibility of a restricted refrigerant flow, which again will result in starvation of liquid to the evaporator.

Other faults, less evident when an engineer is in attendance, can come to light after the repair: refrigerant leakage from a shaft seal may be due to another component failure; an inefficient compressor can lead to loss of duty and extended running time, with unnecessary expense to the customer.

Checking a compressor operation is simple, but will entail adjustment of some controls (see page 51).

The various faults and symptoms are dealt with in this chapter.

Procedure

It is recommended that a quick acting indicating thermometer be used. Two thermometer probes must be firmly attached to pipework free of frost or ice to ensure good thermal conductivity, one at the evaporator inlet and the other as close as possible to the thermal element or bulb of the expansion valve (see

Figure 18). The temperature of the refrigerant passing through the pipework at these two points can now be quickly and accurately taken.

The temperature difference between the two points is the amount of super-heating taking place within the evaporator. A very wide differential means that the coil is starved of liquid and subject to excessive superheating, and that the evaporator is not fully active. Very little or no temperature difference means very little superheating; the evaporator is fully active over its entire surface, but some frosting back could occur at certain times during the system operation.

It is important that the expansion valve is adjusted to give the correct amount of superheating during commissioning, when a replacement has been made and, as good insurance, when a system has been charged.

Thermal bulb

Generally the performance of the thermostatic expansion valve depends upon the correct location of the thermal bulb. This should be clamped firmly to a horizontal section of the suction line near to the evaporator outlet if possible, and preferably within the refrigerated space. The entire length of the bulb should make good thermal contact with the pipework, as illustrated in Figure 18.

The location of the bulb relative to the diameter of the tubing is also important. It must never be located on the underside of the pipework, where it will register false temperatures because of the presence of oil in the pipework.

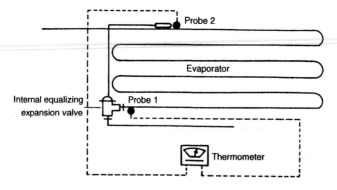

Figure 18 *Expansion valve and probes*

Valve settings

Manufacturers will stipulate the setting of new expansion valves. They are generally set to maintain a superheat of 3.5 to 5 °C or 7 to 9 °F.

Different evaporators require different settings of the valve to keep the coils correctly flooded. It is accepted that optimum operating conditions are achieved when thermostatic expansion valves are set to within the following:
Dry expansion, gravity coil: 5 to 7 °C or 9 to 12 °F.
Forced air evaporator coil: 1.5 to 3 °C or 3 to 5 °F.
It is evident that if a factory set valve with a 7 °C setting is fitted to a forced air evaporator, too much superheating will take place. By adjusting the valve to within the accepted range, say 4 °C, less superheating will occur and the evaporator will be more efficient.

Figure 19 shows two conditions of the refrigerant in a gravity coil at different settings of the expansion valve.

Pressure drop: external equalizing

A low evaporating pressure can be due to pressure drop through the evaporator caused by friction, the length of the evaporator tubing, and the number of return bends employed in the evaporator design. If this occurs, the saturation temperature of the refrigerant may be lower at the outlet of the evaporator than at the inlet.

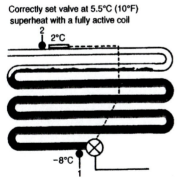

Correctly set valve at 5.5°C (10°F) superheat with a fully active coil
2°C
−8°C
Probe 1 shows inlet at −8°C
Probe 2 shows outlet at 2°C: 10°C superheat
Evaporator fully active

Incorrectly set valve producing 20°C superheat with part of the coil inactive
12°C
−8°C
Probe 1 shows inlet at −8°C
Probe 2 shows outlet at 12°C: 20°C superheating
Evaporator partially active

Figure 19 *Expansion valve settings*

When an evaporator has a pressure drop in excess of 5 psig or 0.3 bar it is essential to employ a thermostatic expansion valve with an external equalizer. This type of valve is designed to compensate for pressure drop. The equalizer is a small tube or capillary which is connected beneath the expansion valve bellows or diaphragm; the other end is installed in the suction line at the outlet of the evaporator. In operation the equalizing tube provides the same pressure as that in the suction line at the thermal bulb location whilst the compressor is running. This equalizing of pressure will permit accurate superheat adjustments.

The expansion valve equalizing connection should be located 150–200 mm beyond the thermal bulb of the valve on the compressor side, as shown in Figure 20.

To further explain the significance of pressure drop and the need for the use of an external equalizing expansion valve, Figure 21 shows the effect of pressure drop with an internal equalizing expansion valve and how external equalizing compensates for the pressure drop. The calculations, pressures and temperatures are for refrigerant R12.

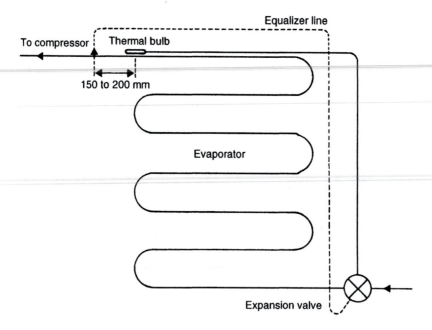

Figure 20   *Expansion valve with equalizer*

Internally equalized expansion valve on
an evaporator with no pressure drop

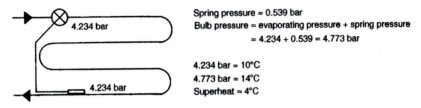

Spring pressure = 0.539 bar
Bulb pressure = evaporating pressure + spring pressure
                    = 4.234 + 0.539 = 4.773 bar

4.234 bar = 10°C
4.773 bar = 14°C
Superheat = 4°C

Internally equalized expansion valve on an evaporator
with a noticeable pressure drop of 0.2 bar

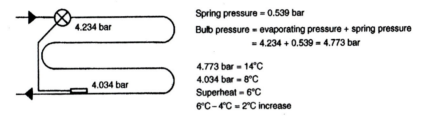

Spring pressure = 0.539 bar

Bulb pressure = evaporating pressure + spring pressure
                    = 4.234 + 0.539 = 4.773 bar

4.773 bar = 14°C
4.034 bar = 8°C
Superheat = 6°C
6°C – 4°C = 2°C increase

Externally equalized expansion valve on an evaporator
with a noticeable pressure drop of 0.2 bar

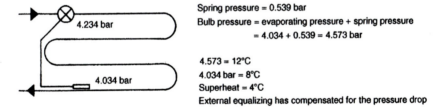

Spring pressure = 0.539 bar
Bulb pressure = evaporating pressure + spring pressure
                    = 4.034 + 0.539 = 4.573 bar

4.573 = 12°C
4.034 bar = 8°C
Superheat = 4°C
External equalizing has compensated for the pressure drop

Figure 21 *Internal and external equalizing expansion valves*

Suction line frosting

When flooding of the evaporator takes place, frosting along the suction line
will be evident. This is caused by the liquid refrigerant boiling off and reducing
the temperature of the suction line. Frosting along the suction line can also
occur when there is a pressure drop through the evaporator and the suction
line temperature is reduced to below 0 °C.

With low temperature applications the returning suction vapour may be at a temperature well below 0 °C, resulting in suction line frosting. It is common practice to insulate suction lines of low temperature installations in order to overcome this problem.

Distributor refrigerant control

Some commercial evaporators are designed with series-parallel and parallel tubing circuits, which means that they have more than one refrigerant circuit. Refrigerant from a single outlet expansion valve is directed to the circuits by means of a distributor, which in most cases is part of the evaporator assembly. These exist in many forms; two types are shown in Figure 22.

A multi-outlet expansion valve can also be used to perform a similar function, feeding two refrigerant circuits. Such a valve is shown in Figure 23.

Replacing a thermostatic expansion valve

There are a number of reasons for replacing an expansion valve:

1 Mechanical failure of the valve: possible seizure of the valve stem carriage; worn valve seat and letting by; defective or wrong sized valve orifice cartridge.
2 Partial loss of the thermal charge: insufficient pressure exerted by the power head in response to temperature changes, resulting in an erratic valve operation.

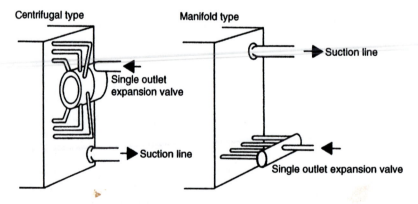

Figure 22   *Distributor refrigerant control*

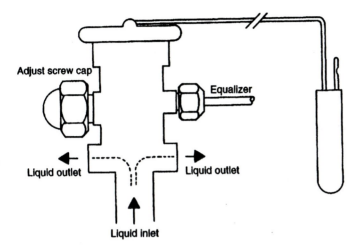

Figure 23   *Double outlet expansion valve*

3   Complete loss of the thermal charge from the power head: the spring pressure will close the valve and refrigerant flow to the evaporator will cease.
4   An excessive pressure drop across the evaporator is discovered: an external equalizing type has to be substituted.

Complete loss of thermal charge

When this condition exists there is effectively a total blockage in the high side of the system (liquid line). The pressure in the low side of the system may be negative, or that determined by the cut-out setting of the low pressure control.

The system is therefore 'pumped down' from the point of the blockage, in this case the expansion valve, and there is no need for the pumping down procedure to be carried out. There is, however, a liquid line full of liquid; unless an isolating valve has been installed, this liquid must be removed.

If there is a liquid charging valve at the receiver with an appropriate connector, it may be possible to transfer the liquid into an empty refrigerant cylinder. If not, it must be reclaimed. The following procedure should be adopted:

1   Isolate the plant electrically and close the liquid shut-off valve at the receiver.
2   Front seat the suction service valve.

3   Release the thermal bulb from the clamp at the outlet of the evaporator.
4   Ensure that the working area is well ventilated. If any food products are in the immediate vicinity they should be removed or covered as a protection against contamination.
5   Check the orifice size of the new valve, and check that it is suitable for use with the system refrigerant.
6   Remove the defective valve and install the replacement. Remove the old bulb clamp and fit the thermal bulb of the replacement valve to the evaporator outlet in the original location, using the new clamp provided with the new valve.
7   Open the liquid shut-off valve at the receiver.
8   Leak test the connections which have been disturbed.
9   Set the suction service valve to the operating position and start the plant.
10   Operate the plant until pressures have stabilized. Check the refrigerant charge via the sight glass and operating pressures.
11   Adjust the valve to give the recommended degree of superheating.
12   Remove gauges, cap service valves and carry out a final leak test.
13   Clear the site of debris, oil etc.

The system may be evacuated through the gauge port on the compressor. Crack off the suction service valve from the front seat position and allow a small amount of vapour to enter the compressor crankcase. Front seat the suction service valve again, slacken off the gauge port union and evacuate the compressor. After evacuating, set the suction service valve to the operating position and continue the above procedure at step 12.

Mechanical failure or partial loss of thermal charge

If the valve is seized in a closed position the procedure will be the same as previously described.

When a valve is letting by or has lost part of its thermal charge, then the system must be pumped down before carrying out the procedure.

Pressure drop

If an internal equalizing expansion valve is substituted with an external equalizing valve, an additional connection must be provided for the external equalizing capillary.

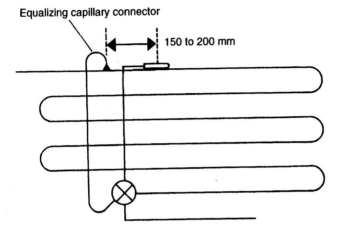

Figure 24 *External equalizing capillary*

The capillary connection must be installed in the suction line as illustrated in Figure 24. The connection is a $\frac{1}{4}$ in flare half union, brazed into the suction line 150 to 200 mm downstream from the expansion valve bulb location.

Replacing the filter drier

Assuming the restriction in the drier is only partial, and that a low pressure cut-out control is in circuit, the following procedure should be adopted:

1 Note the pressure at which the compressor cuts out. The control must be reset to the original setting after the filter drier is replaced.
2 Alter the range of the low pressure control and pump down the system as previously described. Ensure that the suction service valve is front seated when this is completed.
3 Isolate the plant electrically.
4 Remove the defective filter drier and replace it. Take care that the new one is installed correctly; note the direction of flow.
5 Open the liquid shut-off valve at the receiver.
6 Leak test the joints which have been disturbed.
7 Loosen the connection on the suction line at the inlet to the compressor and allow a small amount of vapour to escape, tighten connection.

8 Set the suction service valve to the operating position and operate the plant.
9 Reset the low pressure control to its original cut-out setting.
10 Remove gauges, fit service valve caps and carry out a final leak test.
11 Clear the site of debris etc.

Should the filter drier be completely blocked, which is improbable, then the replacement procedure will be the same as that given earlier for the replacement of a thermostatic expansion valve with a complete loss of charge.

The filter drier is described further in Chapter 12.

Low side purging

Low side purging is now considered to be bad practice and air in the system should be removed by using a vacuum pump to minimize refrigerant loss to the atmosphere which in turn affects the ozone layer. The correct procedure to be adopted is given in the section dealing with reclaim and recovery of refrigerants.

Moisture in the system

When a system pipework is opened to atmosphere during a replacement operation, it is possible that air will enter the system. Air contains moisture, and only a small amount of moisture in a system which has a capillary for the refrigerant control can result in that moisture freezing. This leads to a complete loss of refrigeration.

Filter driers are normally capable of dealing with small quantities of moisture. However, it is recommended that a drier be changed when leaks are detected, especially on the low side.

A complete loss of the refrigerant charge owing to leakage can result in a compressor operating on vacuum, drawing in air. In this case the drier could become saturated and moisture will circulate through it. When this occurs, freezing can take place at the expansion valve; this is tantamount to a complete blockage in the liquid line.

When the symptoms of plant operation indicate a complete blockage, and no temperature difference is obvious at the filter drier, it is natural to suspect that the expansion valve or perhaps a solenoid valve is at fault. The simple expedient of applying a cloth dipped in hot water will determine the presence of moisture. Warming up the expansion valve will melt the ice in the valve and the flow of refrigerant will resume, but only until such time as the temperature at the expansion valve is low enough to form ice and restrict the liquid flow once more. A blowtorch should never be used for this purpose, for obvious reasons.

In most cases, fitting a new filter drier will overcome the problem. However, it is stressed that this drier should be removed after a suitable running period and another new drier fitted. Should the condition persist after a drier has been replaced, the system must be discharged of refrigerant, evacuated and then recharged.

Compressor efficiency test

This is also referred to as a compressor pump test. It is carried out when the functions of the compressor suction and discharge valves are suspect; this will be indicated by high suction and low discharge pressures. The test should be carried out with the plant running at an operating head pressure of at least 100 psig or 6.5 bar if possible in order to prove the efficiency of the valves.

The procedure is as follows:

1 Front seat the suction service valve and note the cut-out pressure of the low pressure control. The control must be reset to its original cut-out pressure after the test.
2 Alter the range of the low pressure control so that the compressor is drawing a vacuum.
3 Reduce the low side pressure to at least 20 in Hg vacuum or 0.7 bar.
4 Stop the compressor and observe the pressure rise on the compound gauge for 2 minutes.

If the suction reeds are seating properly and cylinder/piston wear is not excessive, the pull-down to a vacuum should be rapid. The front seating of the suction service valve isolates pressure coming from the evaporator, and only the compressor crankcase is being evacuated of the refrigerant vapour. A compressor is deemed reasonably efficient if the 20 in vacuum is achieved. If it is not possible to draw this vacuum, then the suction reeds are defective.

When a 20 in vacuum is achieved, the pressure rise should be minimal; the entire vacuum should not be lost over the 2 minute observation period. When the vacuum is drawn and the plant switched off, if a rapid rise in pressure is observed on the compound gauge then refrigerant vapour is leaking into the cylinder(s) via the discharge valve(s). The valves must be inspected for distortion of reeds or faulty seating.

Removing compressor valve plate assembly

On most compressors this will be a relatively simple task, but the procedure will differ with compressor design. The following is the procedure for a compressor with the valves in the head:

1 Start the plant, front seat the suction service valve and reduce the crankcase pressure to 3 psig or 0.2 bar if possible.
2 Stop the plant, isolate electrically and front seat the discharge service valve.
3 Slacken off the compressor head bolts slowly to release the high pressure from the discharge side of the head. Remove the bolts.
4 Gently raise the compressor head, with suction and discharge lines intact, sufficiently high to be able to withdraw the valve plate assembly. Care must be taken to avoid fracture of the pipework unions.
5 If the cylinder head and valve plate gaskets are damaged when the valve plate is removed, they must be replaced.
6 Distorted or broken reeds will obviously have to be replaced. Some manufacturers will recommend replacement of the entire valve plate assembly; replacement kits are available complete with gaskets.

Should the valve seats on the valve plate be eroded or pitted and a replacement is not readily available, they can be made serviceable by reseating or lapping. This is generally regarded as a workshop practice. The method is as follows:

1 Obtain some valve grinding paste or carborundum powder and some polishing compound. Spread a liberal amount of the paste on to a lapping block or a hard flat surface such as a polished steel plate or sheet of glass.
2 Remove the discharge reed retainers and reeds from the valve plate.
3 Place the valve plate on the pasted surface, exert a firm even pressure and move the valve plate in a figure-of-eight motion. Continue until the valve seats are returned to an original finish. When carborundum powder is used it should be mixed with refrigeration oil.
4 When reseating is complete, remove all traces of carborundum and paste from the valve plate with spirit or paraffin.
5 Repeat the operation with the polishing compound and clean as before.
6 Reassemble dry with new reeds if required. If the original reeds are used, they should be inverted so that the seat contact is made to the unused side of the reed.

The procedure for replacing the valve plate assembly is as follows. If gaskets are re-used they should be perfectly clean and dry.

1 Once the valve plate is located, replace the head bolts and screw them down finger tight.
2 Tighten the bolts diagonally across the head, care being taken not to overtighten and strip the threads.
3 Crack off the suction and discharge service valves from the front seat positions.
4 Leak test the compressor and pipework unions.
5 Purge the compressor through the gauge port unions.
6 Set the service valves to the operating positions and start the plant.
7 Operate the plant and observe pressures; these should now be normal.
8 Carry out the compressor pump test.
9 Reset the low pressure control.
10 Remove gauges, replace valve seal caps and wipe the compressor clean of oil.
11 Make a final leak test.
12 Clear the site.

Removing a compressor rotary shaft seal

Most smaller open drive and some direct drive semi-hermetic compressors employ this type of crankshaft seal. They are often a cause of noise complaints (squeaking) and refrigerant leaks. Generally this is due to lack of lubrication; when the seal surfaces are dry, wear and scoring of the polished facings occur.

In order to remove this type of seal it is necessary first to remove the compressor flywheel. The larger the compressor, the larger and heavier will be the flywheel. Extra care must be exercised when removing the larger types, which may be castings and easily damaged if dropped. Removal should not be attempted without a suitable extractor. Under no circumstances should a hammer be used to break the bond between shaft, keyway and flywheel boss (not for semi-hermetic compressors).

Systems must be pumped down and isolated electrically. The drive belt guards must be removed to gain access to the flywheel on units with water cooled systems or with a remote condenser. On smaller air cooled units the compressor must be removed from its base, and this is dealt with here.

Open drive units (valves in the head type)

1 Pump down the system, front seat both service valves and isolate the unit electrically.

2   Remove the head retaining bolts and gently raise the compressor head.
3   Withdraw the valve plate assembly and remove the suction reeds from the cylinders.
4   Remove the compressor mounting bolts and belt guard.
5   Slide the compressor body towards the drive motor, release the drive belts from the flywheel and remove the compressor body.
6   Release the locking device on the drive shaft. This could be in the form of two locknuts, a locknut and a tab washer, or a locknut and a pin.
7   Using a suitable extractor, locate the arms around the flywheel boss (avoid locating around the flywheel vee section). Never use a hammer to remove a flywheel.
8   Apply gentle pressure on the extractor to break the bond, then remove the flywheel. With large heavy types it is advisable to tie a rope or cord to the flywheel and secure it in case the bond breaks suddenly.
9   Remove the seal plate retainer bolts and seal plate. The seal will be released by spring pressure in most cases; withdraw the seal from the seal housing.
10   Remove the seal ring from the shaft.

It may be necessary to change the compressor oil when a seal replacement is made, or some oil may be lost during replacement.

The procedure for replacement is as follows. Some manufacturers supply a shaft centring tool to ensure that the seal is correctly aligned on the shaft; this also eliminates uneven pressure being applied by the seal plate and spring during assembly.

1   Locate the seal ring on to the shaft and push fully home to the shaft shoulder.
2   Dip the seal nose in refrigeration oil and locate over the shaft, using the centring tool if available.
3   Replace the seal plate, depressing evenly against the spring. Insert the retaining bolts, and tighten the bolts so as to maintain the correct alignment of the shaft seal assembly.
4   Reassemble the compressor, belts etc. in reverse order. Check the belt tensioning and the alignment of the drive option and flywheel.
5   When reassembly is complete, fit gauges if these have been removed to facilitate the replacement.
6   Crack off both service valves from the front seat positions and purge the compressor through the gauge port unions if there is sufficient refrigerant pressure in the system. If the charge was lost, then evacuate, fit a new filter drier, charge the system and check the oil level.
7   Carry out a leak test.

8   Start the plant and check the system operation.
9   Remove gauges.
10 Clean the compressor and clear the site.

Direct drive units

This type of compressor does not have a flywheel or drive belts; the compressor is linked to the drive motor by means of a coupling. To gain access to the seal housing, this will have to be dismantled before the seal plate and seal assembly can be withdrawn. It may also be necessary to move the drive motor in order to remove the seal assembly.

Basically the foregoing procedures for removal and replacement remain the same, with the exception of the coupling alignment. A typical alignment is dealt with in Chapter 13.

Excessive operating head pressure

This condition is probably more common than faults causing low suction pressures, especially during the summer months when ambient temperatures are higher. Causes and corresponding remedial actions are listed as follows.

Air cooled systems

Causes

Restricted air flow over the condenser is caused by:

1   Condenser fins blocked by an accumulation of dirt and debris drawn in by the condenser fan(s).
2   Inoperative condenser fan(s).
3   Air in the system if a leak has developed on the low side and the compressor has operated with suction pressures below atmospheric.
4   High ambient temperature.
5   An overcharge of refrigerant.

Remedies

1   The most effective method of cleaning condensers is by means of a liquid or foam application which penetrates the build-up of the undesirable deposits

on the coil and fin surfaces. Brushing the condenser can produce somewhat limited results because of inability to reach the entire surface area.

2 Replacement of fan(s) may be required, or failure may be due to a loose wire or broken electrical lead.

3 This can be verified by stopping the plant and allowing sufficient time for the condenser to cool to ambient temperature. Then refer to a pressure/temperature table and compare the standing or idle pressure with the pressure given by the table. If the idle pressure is higher than that given by the table, air or non-condensables are present in the system. Running condenser fans can speed up the cooling process.

4 High ambient conditions will require a survey of the installation and location of the condenser. Relocation may overcome the problem and provide larger volumes of fresh air. An extractor fan could be installed to remove the discharged air from the condenser, so preventing recirculation. When multiple units are installed, a baffle arrangement to route a fresh air supply over each condenser must be considered.

5 An overcharge of refrigerant cannot develop and must be the fault of the service or installing engineer. Check the standing or idle head pressure in the same manner as for air in the system. The pressure/temperature relationship should conform to the table. The excess refrigerant must be removed from the system.

Water cooled systems

Causes

Restricted water flow through the condenser is caused by:

1 Scaling of the interior surface of the condenser water tubing.
2 Incorrectly adjusted or defective water regulating valve.
3 Inadequate water supply: malfunction of recirculating water pump, resulting in poor supply and high water temperature.
4 Air in the system.
5 An overcharge of refrigerant.

Remedies

1 Descaling of condenser tubing can be carried out by brushing through the tubes of a shell and tube condenser. If the scale deposit is heavy, a chemical method is advisable. A shell and coil condenser can only be cleaned chemically. The cost of descaling must be compared with that of

a replacement. The cleaning of water cooled condensers is described later in this chapter.

2 Check the water inlet and outlet temperatures of the condenser and the water regulating valve operation.

3 Check the water supply pressure and volume.

4 Air or non-condensables present in a water cooled condenser system can be diagnosed very quickly. The water regulating valve, responding to the high operating head pressure, will be supplying a high volume of water to the condenser. When the plant is stopped, water will continue to flow and reduce the temperature of the refrigerant in the condenser. In a matter of minutes the temperature of the refrigerant will be the same as that of the water (the inlet and outlet temperatures will be equal). Compare the idle head pressure with that given in a pressure/temperature table; a higher than normal pressure denotes the presence of non-condensables.

5 Adopt the same procedure as in remedy 4.

If high operating head pressure conditions cannot be rectified by any of the foregoing, it must be considered that the condenser is undersized. It must be replaced, or a subcooler must be installed.

High side purging

Like low side purging, the practice of purging from the high side is also unacceptable and **must be discontinued**. When it is certain that non-condensables are present in a system a leak test must be carried out on the entire system and the leak repaired. If a plant has been recently installed or it has been subject to service or repair it is possible that nitrogen could have been inadvertently left in the system. If air has entered the system in sufficient quantity the effect could possibly be of excessive operating head pressure similar to that of nitrogen left in the system. The removal of a non-condensable can be costly if the operating charge of refrigerant is large. The system would need to be completely discharged, evacuated and then re-charged.

A non-condensable gas will be lighter than refrigerant vapour and will rise to the top of a condenser/receiver when the plant is at rest. Removal of the non-condensable must take place at the highest point. Some condensers/receivers are fitted with a gauge connector or valve for this purpose. This connector may be in the form of a 'purge screw' on older type equipment. Removal of a non-condensable can easily be achieved via the gauge port on the high side of the compressor head. The system should be 'pumped down' and a short period of time allowed for the non-condensable to rise to the highest point. The procedure for removal is detailed in the section covering recovery/removal

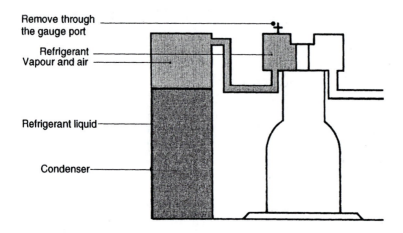

Remove through
the gauge port

Refrigerant
Vapour and air

Refrigerant liquid

Condenser

Figure 25   *High side purging*

of refrigerant. Figure 25 shows the location of the refrigerant and non-condensable (air) in the high side of a system and the gauge connection.

The removal of non-condensables therefore should be carried out by using a vacuum pump. Purging air from a system charged with refrigerant would also release some of the refrigerant to the atmosphere.

It is the duty of any service or installation engineer to protect the environment by discontinuing the release of CFCs and HCFCs to the atmosphere whenever servicing or installing refrigeration equipment.

In the case of nitrogen being present in a system, this can be purged from both the low and high side.

Water cooled condensers: scale and corrosion

Scaling

Any type of water cooled condenser is prone to scaling of the interior surfaces of the water tubing. The rate at which the scale forms will depend upon the condensing temperature and the quality of the water circulated. Scaling will be relatively low where condensing temperatures are below 38 °C.

Shell and tube condensers are fitted with removable end plates to enable cleaning to take place by means of a wire brush. This is satisfactory for mild scale build-up, but heavy scale deposits may necessitate removal by chemical means.

Scale can be removed with descaling agents, which are available in both liquid and powder forms. Solutions of chemicals and water can be used as follows:

1 Muriatic acid 18 per cent and water 82 per cent for rapid descaling.
2 Hydrochloric acid 22 per cent and water 78 per cent for slower but equally effective action.

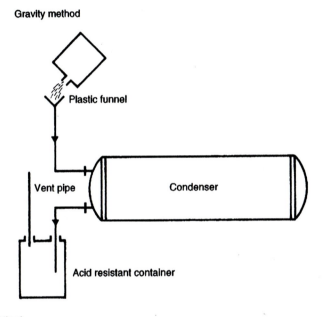

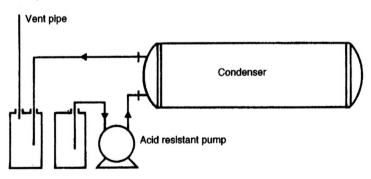

Figure 26 *Condenser descaling*

Descaling processes can be carried out by a gravity flow method or by means of a pump. However, it must be realized that all descaling agents are acid based; therefore any pump must be acid resistant. Never use a system recirculating pump for this purpose. Figure 26 shows layouts for the two methods.

Descaling solutions can damage flooring, paintwork, clothing and plant life. They are obviously a health hazard, and contact with eyes and skin must be avoided. Adequate protective clothing must be worn, and precautions against spillage must always be taken.

When cleaning condenser coils or tubes the working areas must be well ventilated. A vent pipe is a vital part of the cleaning equipment; it will carry off the toxic fumes which are generated by the chemicals during the cleaning process.

Corrosion and contamination

In addition to scaling, water cooling systems are subject to corrosion from fumes given off by nearby industrial plant. Concentrations of sulphur and salts will be present in the atmosphere.

Systems installed in a coastal area are susceptible to corrosion by salt borne by the air.

There will always be the problem of algae growth and bacterial slime. These can only be controlled by regular cleaning and the use of various algaecides. Inhibitors such as Hydrofene have been used over a long period without any ill effect being observed.

Compressor motor burn-out: system flushing

Prolonged operation at high discharge pressures and temperatures, excessive motor starting, fluctuating voltage conditions, shortage of refrigerant charge and shortage of oil in the compressor are all possible causes of a motor burn-out.

A burn-out can be defined as the motor winding insulation having been exposed to a critical temperature for a long period.

Refrigerant thermal decomposition

This occurs with R12, R22 and R502 at temperatures in excess of 150 °C (302 °F). In the presence of hydrogen-containing molecules, thermal decomposition produces hydrochloric and hydrofluoric acids. Phosgene is produced at very high temperatures, but this is decomposed in the presence of oxygen.

Bearing in mind the above, it is important to remove the entire refrigerant charge and reclaim the refrigerant for processing. A recognized reclaim refrigerant cylinder should be used and care taken not to overfill the cylinder.

Acid testing

When the windings insulation breaks down, very high temperatures occur at the short circuited location. In addition, a certain amount of moisture will be released from the windings assembly to further contaminate the system. Following a motor burn-out, the system must be decontaminated before a new compressor is fitted.

When the defective motor has been removed, a test should be made to determine the acid content of the compressor oil. Two methods may be used: litmus paper and burn-out test indicator. A sample of the oil from the defective motor compressor should be taken and tested. If the test indicates acid, then the refrigerant system must be flushed and tested as follows. Flushing is dealt with overleaf.

Litmus paper

Take a sample of the solvent after flushing and put it into a suitable container. Place the litmus paper in the liquid. If acid is present it will change colour, ranging from pink to red according to the degree of acidity in the sample. The system must then be flushed again and the test repeated.

Indicator

When testing with an indicator it is necessary to charge the system with the liquid solvent ready for flushing and allow to stand for 30 minutes. Then take samples of the solvent, if possible from both the high and the low side of the system.

Add the prescribed amount of indicator to the samples and agitate the mixture; examine for a colour change. The results and necessary actions are as follows:

1 If red or pink, strong acid content: flush again.
2 If orange or yellow, acid content: flush again.
3 If carbonized particles are present in the samples: flush again.
4 If lemon yellow, no acid content: system may be evacuated.

System flushing

When a system has been contaminated, especially following a hermetic motor compressor 'burn out', the past practice was to flush the system through with R11. **This practice is no longer acceptable**.

Approved burn-out filter driers are available these days to make flushing unnecessary. Instead the system can be cleansed by installing suitable filter driers and carrying out a triple evacuation (see Chapter 16, Dilution method). The filter driers will absorb moisture and acid content from the system pipework.

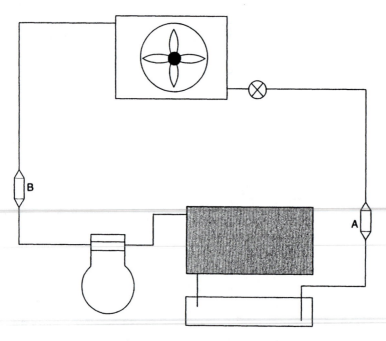

A = Burn-out drier installed in the liquid line
B = Burn-out drier installed in the suction line
(optional). This is sometimes advisable
when a system is contaminated as a result
of a hermetic or semi-hermetic motor
compressor burn-out

Figure 27   *Evacuation method*

Evacuation

During the evacuation of the system, evaporator fans and electric defrost heaters may be switched on to raise the temperature of the evaporator. However, extreme care must be taken to avoid overheating by the defrost heaters.

Burn-out drier

After a system has been repaired and evacuated, burn-out driers installed, leak tested and charged with refrigerant it should be operated for a period of 24 hours. An acid test should then be carried out and, if satisfactory, the burn-out driers can be removed and exchanged for normal filter driers. When an acid test reveals contamination new, burn-out driers must be installed and the process repeated.

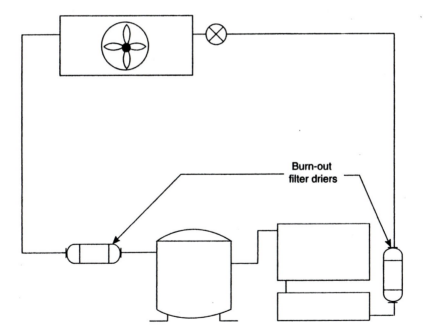

Figure 28 *Burn-out driers installed*

Pressure controls

There are many different makes of pressure controls. Designs may vary but the principle of operation is basically the same: single pressure models respond to either low or high pressure; dual pressure versions can be activated by both high and low pressures.

Pressure controls are installed to perform a variety of functions. Most current production controls are designed with three electrical terminals. These enable the control to be used to open an electrical circuit upon a rise in pressure or upon a fall in pressure, depending on the application. Unfortunately, identification of the electrical terminals is not standard, and it is important to know which of the three is the common terminal.

Single pressure controls have two adjusting screws: the range and the differential. Dual pressure versions have three screws: the range and differential for low pressure operation, and a range screw for high pressure operation. This differential is preset at 50 psi or 3.5 bar.

The controls described are commonly employed and will serve to identify adjusting screws and electrical terminals.

Applications

A low pressure control may be used as:

1   A temperature control.
2   A safety cut-out to prevent overcooling or overfreezing of a product.
3   A defrost termination control to open the circuit to defrost heaters should the evaporator become completely free of frost before the defrost period has elapsed. This prevents undue pressure build-up in the evaporator.
4   A compressor cut-out when operating on a pump-down cycle.

Table 4   *High pressure control settings*

| Refrigerant | psig | | bar | |
	Cut-out	Cut-in	Cut-out	Cut-in
12	210	160	14	12
22	290	240	19	16
502	310	260	20.5	17.5

5 An evaporator fan delay control to prevent warm air and moisture from being circulated to the refrigerated space after defrost.

A high pressure control is employed as a safety cut-out for the compressor in the event of excessive operating pressures developing on the high side of the system during operation. Table 4 shows the settings required.

Example controls

Figure 29 shows two typical controls. The details are as follows.

Ranco type 016: low pressure operation

To complete an electrical circuit on rise in pressure, terminals 1 and 4 must be used. To complete an electrical circuit on fall in pressure, terminals 1 and 2 must be used.

The control range is 12 in Hg vacuum to 100 psi (0.4 to 7 bar).

Ranco type 017: dual pressure operation

The control range is 12 in Hg vacuum to 100 psi (0.4 to 7 bar) on the low pressure side, and 100 to 400 psi (7 to 26.5 bar) on the high pressure side. The high pressure automatic reset differential is 50 psi (3.5 bar).

Control setting procedure

For low pressure control:

1 Adjust the range screw A until the compressor cuts in at the selected pressure.
2 Adjust the differential screw B to stop the compressor at the desired cut-out pressure.

For high pressure control, adjust the range screw H to a selected cut-out pressure and the control will reset automatically.

Check the control operation several times after final adjustments have been made.

Always adjust controls with a pressure gauge, using the graduated scales as a guide only. Never lever the control mechanism to actuate it; this can

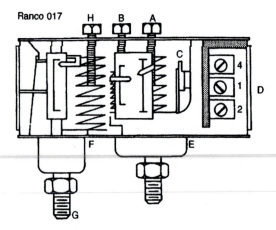

Ranco 016

Ranco 017

A range adjusting screw   E low pressure diaphragm
B differential adjusting screw F high pressure diaphragm
C toggle lever   G flare connection
D electrical terminals   H high pressure range adjusting screw

Figure 29 *Pressure controls*

damage the pivot and cause erratic operation or failure. Use the insulated lever if fitted.

This procedure can be followed for setting any make of control, such as those in Figure 30.

When locking plates are provided, set the control, replace the locking plate to secure the differential screw, and fit the knob to the range screw.

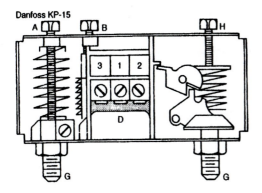

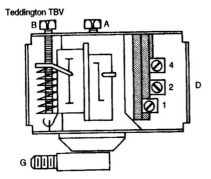

KP-15:   set cut-in on start scale (low pressure)
             set cut-out on stop scale (high pressure)
             for break on fall in pressure, use terminals 2 and 3
             for break on rise in pressure, use terminals 1 and 2

TBV: for make on rise in pressure, use terminals 1 and 4
           for break on rise in pressure, use terminals 1 and 2

Figure 30   *Pressure controls*

Pressure adjustments

Low pressure control adjustments can be made by front seating the suction service valve to reduce the pressure on the low side of the system.

High pressure control adjustments must be made by increasing the operating head pressure, i.e. by blocking off the condenser, thereby restricting the air flow, or by switching off condenser-fans. The head pressure can be quickly increased on water cooled units by shutting off the water supply.

It is dangerous practice to increase the head pressure by front seating the discharge service valve.

Control examples

The examples in Figure 31 illustrate the effects of a reduced range and differential setting with a low pressure control.

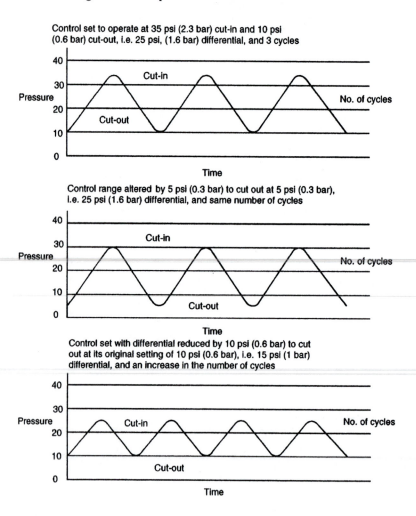

Figure 31   *Control samples*

Motor cycling controls: thermostats

Two types of thermostat are commonly used to control the temperature of a refrigerated space or product by stopping and starting the compressor.

The most popular is the vapour pressure type. This consists of a small bulb or sensing element containing a very volatile liquid charge. The liquid has the ability to vaporize at low temperatures. When the bulb is subjected to a rise in temperature, the pressure generated by the vaporizing liquid will increase.

A capillary connects the bulb to a bellows or diaphragm. The vapour pressure causes the bellows to expand or the diaphragm to flex and operate the switch mechanism, closing the contacts to start the compressor motor. As the sensing element or bulb cools, the pressure in the bellows or diaphragm will decrease. The bellows will contract or the diaphragm will return to its normal position, thus opening the electrical contacts to stop the compressor.

Switch mechanisms have some form of toggle action or a permanent magnet device to ensure rapid and positive make or break of the switch contacts. This prevents arcing, which occurs when electrical current jumps across the minute gap between the contacts. Figure 32 shows toggle and permanent magnet switch arrangements.

The permanent magnet snap action type has a switch contact arm made of a magnetic material (iron or steel), and the magnet attracts the arm towards it. The pressure from the sensing element acts against the magnetic attraction to close the contacts according to the temperature of the element or bulb. As the arm moves closer to the magnet, the magnetic effect increases to cause the snap action.

When the sensing element cools and the switch bellows contract, some force is necessary to open the contacts. However, the magnetic force decreases as soon as the contacts break to allow rapid opening.

The compound bar type of thermostat, more generally called a bimetal element, comprises two dissimilar metals, usually Invar and brass or Invar and steel. Invar is an alloy which has a very low coefficient of expansion, whilst brass and steel have a relatively high coefficient of expansion. When an increase or decrease of temperature is sensed by the bimetal element the length of the Invar will cause the bimetal to warp. This warping action is utilized to open and close the switch contact (Figure 33).

Thermostats have range and differential adjustments which can be altered in the same manner as previously described for pressure controls. Electrical terminal arrangements will obviously differ by manufacturer, and reference should be made to the literature supplied with the controls.

Toggle arrangement

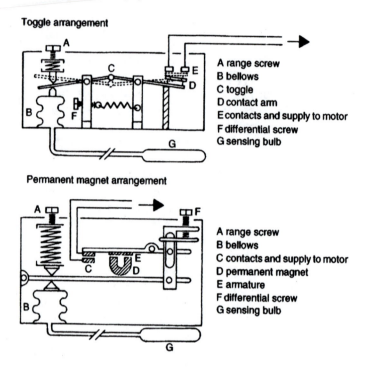

A range screw
B bellows
C toggle
D contact arm
E contacts and supply to motor
F differential screw
G sensing bulb

Permanent magnet arrangement

A range screw
B bellows
C contacts and supply to motor
D permanent magnet
E armature
F differential screw
G sensing bulb

Figure 32 *Vapour pressure thermostats*

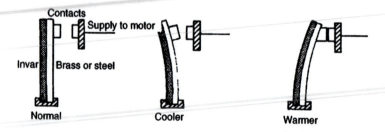

Figure 33 *Bimetal thermostat*

The range on thermostats used on domestic appliances is altered by turning the cold control. Some of these are marked with a warmer or cooler setting. Others are numerical: the higher the number, the lower the temperature. Adjustment of the differential should not be attempted.

Some domestic appliance thermostats have a two-way switching feature in their design. When cooling is no longer required and the cut-out point of the switch is reached, the contacts to the compressor are opened and the contacts to a defrost heater are closed. This achieves automatic defrosting after every on cycle, to keep a minimal frost build-up on the evaporator at all times for more efficient operation.

System faults

Sealed systems such as domestic refrigerators and freezers need to be treated differently, since there may not be provision for fitting gauges unless line tap valves are used.

If the temperature control is by thermostat only and there is no low pressure safety control, continuous running of the unit will result if the system:

1 Is short of refrigerant or has a restriction in the liquid line.
2 Is operating in a high ambient temperature or with restricted air flow over the condenser.
3 Has a faulty fixture door seal, imposing a high evaporator load.
4 Has excessive frost build-up on the evaporator.
5 Is operating with a defrost heater in circuit (has a defective timer), giving a high evaporator load.

Domestic system repairs will be dealt with in detail in Chapter 8.

Noise

The sources of noise are numerous, and in some cases are entirely dependent upon the location and the type of building structure.

Common sources of noise are as follows:

1 Line rattles, that is pipework or control tubing vibrating against an evaporator or condenser shroud or framework.
2 Items of stored products (metal containers) or storage shelves vibrating.
3 Evaporator or condenser fan mountings loose.
4 Condensing unit base mountings or mounting frame loose; inadequate noise suppression from mounting fabric used (Tico pads, for example).
5 Perished rubber mounting grommets, which are especially prone to deterioration if contaminated with oil and refrigerant.
6 Gas pulsations against a rigid pipework design, creating a noise source during operation or when the compressor starts and stops.
7 Incorrect belt tension or belt and drive pulley alignment with open-type condensing units.

A better assessment of noise problems can be made after a study of the pipework design and installation recommendations (Chapter 10).

Seven simple steps for service diagnosis

In order to diagnose any refrigeration fault quickly and accurately, a set procedure must be followed. The procedure described here takes the form of seven simple steps. If fully understood, these will prevent expensive call-backs and dissatisfied customers. No attempt to correct a condition should be made until the fault has been found, and therefore a thorough diagnosis is essential.

It is necessary to have the correct tools and instruments with which to carry out the procedure. These include a resistance thermometer, a leak detector, gauges, valve keys, a multitester or avometer and a compressor test cord, together with a complete set of engineering tools.

The seven steps are as follows:

1 Check the actual temperature of the product and compare with that recommended for the product.
2 Check the suction pressure, control switch settings and product classification to establish the temperature difference (TD) between the evaporator and the product.
3 Check the superheat setting of the expansion valve.
4 Check the condensing medium temperature.
5 Check the operating and idle head pressures of the compressor.
6 Check the refrigerant type and charge.
7 Check the drive pulley size on an open type system. If the compressor is hermetic or semi-hermetic, check the operating range; they may be for high, medium or low back pressure operation.

An incorrect size of pulley may be fitted to a drive motor. The refrigerating effect may be acceptable in cooler ambient temperatures, but when ambient temperatures rise the equipment will not have the capacity because of the compressor speed.

The same principle applies to the compressor operating range; older models were selected for specific operating conditions.

The following practical sequence is suggested for covering the first six steps:

(a) Ensure that the product has been stored for sufficient time to have become chilled or frozen (has not recently been deposited) before checking the product temperature. Take the actual temperature of the product and not the air circulating around it. Take care that the temperature has not been affected by the opening of the fixture door.

(b) Fit gauges and calculate the average suction pressure to establish the TD (see Chapter 5). Consult the classification in Chapter 5 to establish the product group, and note the TD for the type of evaporator employed in the system. If the TD is incorrect, an adjustment of the temperature control may be all that is necessary to rectify the fault.

(c) Ensure that the system has been operating for a sufficient period to be at an average suction pressure, and check the refrigerant charge. Check the operating head pressure and the idle head pressure. Check the superheat setting of the expansion valve. When checking the refrigerant charge, observe the condition of the evaporator (should be fully frosted) and the liquid sight glass. Do not adjust the expansion valve unless the refrigerant charge is complete and the system is operating at an average suction pressure.

7 Electrical equipment and service

Description

This chapter is devoted to the electrical aspects of hermetic and semi-hermetic motor compressors and remote-drive motors employed in commercial refrigeration and in domestic refrigerators and freezers for single-phase 240 volt 50 hertz supply.

The smaller hermetic compressors are mainly used for domestic systems. Larger capacity units are to be found in coldrooms, refrigerated display counters and display cases. These compressors are supplied in various sizes and designs; some operate with a reciprocating action, others are rotary types.

The starting devices for these single-phase compressors are described and their operation is explained, with test procedures which are the same for both hermetic and semi-hermetic compressors.

Open-type reciprocating compressors are also employed for the commercial range. The starting device for remote single-phase drive motors is included, since the test procedures are again basically the same.

Drive motors and starter contactors for both single and three-phase supply are dealt with in Chapter 15.

The split phase motors employed can be capacitor start for low starting torque, and capacitor start and run for high starting torque, on a 200/240 volts 50 hertz supply.

Different types of starting device are used: the current relay for low starting torque, and the potential (voltage) relay for high starting torque. A form of solid state device may be used for either. Regardless of type, each will have overload protection. There will be three external terminals to which the compressor controls are connected to complete the circuit to the two motor windings. The three compressor terminals are known as common, start and run (C, S and R).

External overload protectors are both current and temperature sensitive; they are easily replaced. Internal overload protectors tend to be less current sensitive and provide better motor protection, but cannot be replaced; this means that correct diagnosis of a fault is imperative.

Of the two motor windings, the main or run winding is the larger in both physical size and cross-section. The other is the start or auxiliary winding. They are both conductors, but current will flow freely through the larger winding and meet resistance in the smaller winding.

Service equipment

To carry out an efficient service and diagnosis, the engineer will require the following instruments: voltmeter, ammeter, ohmmeter (or a multitester), wattmeter, test cord and (not essential) capacitor tester.

The wattmeter is used in conjunction with the manufacturer's data or that given on the compressor nameplate. Power readings are an aid to diagnosing faults in much the same way as operating head pressures are for open-type compressors. For example, the average running watts are generally provided on the nameplate or model identification plate. Manufacturers' service manuals will give starting watts, running watts and current consumption under specified conditions for each model.

Current relay

This can best be described as a magnetic switch. It comprises a small solenoid coil around a sleeve and an iron core. Inside the sleeve is a plunger to which the switch contact bridge is attached; the contacts are normally open.

When the coil is energized, a strong magnetic field of force is created because the current will be high during the starting phase. The magnetic force will move the plunger upwards and bridge the switch contacts, completing the circuit to the start winding. The run winding is wired through the relay so that it is always in circuit. Figure 34 shows a typical motor and control arrangement for a current relay, and Figure 35 shows the completed relay circuit.

A high starting current is drawn when the compressor motor starts. The current reduces as the motor gathers speed; the magnetic field through the relay then becomes weaker so that it can no longer hold the contact bridge on to the switch contacts. The plunger then drops down by gravity to open the circuit to the start winding.

It is not uncommon for a start capacitor (see later) to be fitted when a current relay is employed. This is wired in series with the start winding (Figure 36).

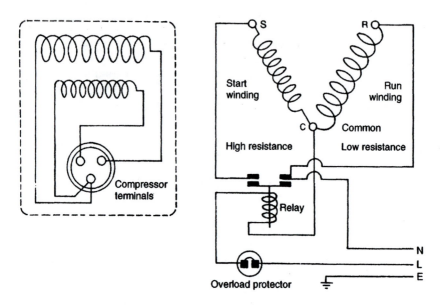

Figure 34 *Current relay arrangement*

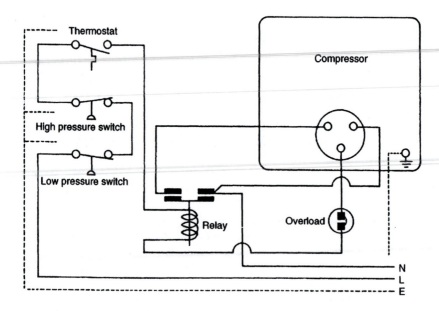

Figure 35 *Current relay circuit*

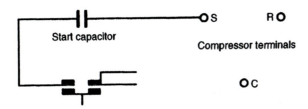

Figure 36 *Start capacitor in circuit*

Potential (voltage) relay

This type of relay is used with high starting torque motors. It operates in a similar manner to the current relay except that the switch contacts are normally closed. The solenoid coil, once energized, maintains a magnetic force strong enough to open the switch contacts and keep them open whilst the compressor is running.

The relay has a much higher design voltage rating than the supply voltage. As the motor approaches its design speed, the voltage across the coil can sometimes be more than twice that of the supply voltage.

When power is supplied to the circuit, the relay contacts are closed. Both motor windings are energized and starting is achieved. As the motor increases speed, the voltage in the start winding increases to cause an increase in both voltage and current passing through the coil. When the design voltage of the coil is reached, the current creates a strong magnetic force to pull in the plunger and contact bridge to open the start circuit, but allows the compressor to operate on the run winding.

When the relay contacts open, the voltage and current across the coil will decrease but will maintain a magnetic force strong enough to keep the contacts open until power is disconnected. The contacts will then return to the closed position ready for a restart.

This relay is wired in parallel, unlike the current type which is wired in series. A voltage relay circuit is shown in Figure 37.

Solid state starting devices

Semiconductor starting devices are used on domestic and commercial units with fractional horsepower motors. Some versions are not recommended for permanent installation but are used rather as emergency replacements, because they may not always be suitable for the motor design characteristics.

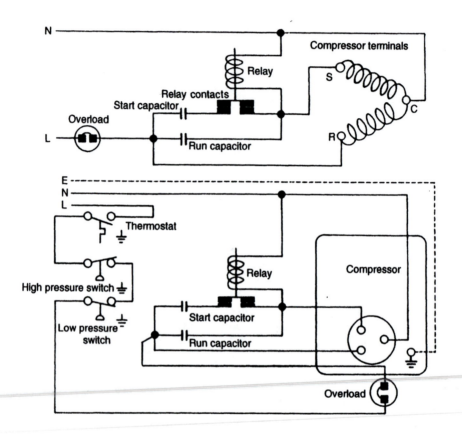

Figure 37   *Voltage relay circuit*

Positive temperature coefficient (PTC) device

Many commercial motor compressors have this type of starting device. In most cases the motor is designed with an internal overload protector; this gives better protection against overload conditions since it is more sensitive to the temperature of the motor windings. It is less current sensitive than the external overload protector.

When the compressor starts, resistance in the semiconductor is low and current can pass freely to the start winding. As the current flows it heats the semiconductor (in approximately 2 to 3 seconds). This causes an increase in resistance through the semiconductor, thereby reducing the flow of current; by this time the motor is up to speed. The reduced current flow is sufficient to maintain the heat in the semiconductor and prevent high current flow to the start winding, allowing the compressor to operate on the run winding.

In the event of a starting failure the increased current drawn will also cause a rise in temperature of the motor windings and actuate the overload protector. Obviously current will not flow to the start winding after a starting failure. The semiconductor has to cool before another starting attempt can be made, and there is therefore greater motor protection.

A further advantage of this type of starting device is that it does not have any moving parts, unlike current and voltage relays. This lessens the risk of failure due to wear and corrosion of contacts. It also makes it more acceptable where stringent demands are made for low noise levels; arcing across switch

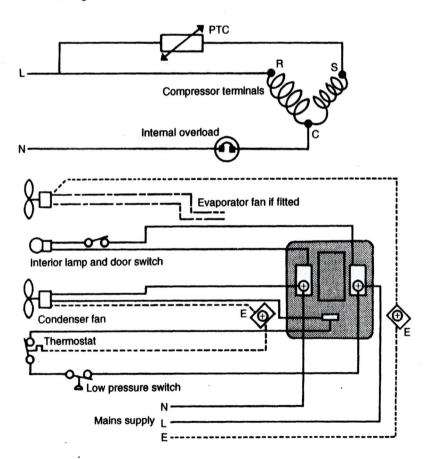

Figure 38 *Basic PTC circuit and wiring diagram*

contacts can affect tape recorders, videos and other domestic and industrial electronic equipment.

The normal cooling period before a restart is 3 to 5 minutes. The protector cut-out temperature is about 140°C or 285°F, and the cut-in temperature is about 105°C or 220°F. At 140°C the cooling period before the compressor can restart may be as long as 45 minutes, depending upon the ambient temperature.

A basic PTC circuit and wiring diagram is shown in Figure 38.

IC-22 relay

This is another form of solid state relay for non-capacitor and capacitor start motors. It can be used as a replacement for the conventional types.

The normal operating temperature of the semiconductor is approximately 76°C or 170°F. Therefore care must be taken when handling if the compressor is running.

Installation for non-capacitor start motors is as follows:

1   Remove the defective relay but leave the overload protector in place if it is serviceable.
2   Connect one lead of the IC relay to the compressor start terminal.
3   Connect temperature/pressure controls, fans etc. to the run lead of the compressor using the quick connector provided.

For capacitor start motors, follow steps 1, 2 and 3, then adapt the compressor wiring to connect the capacitor in series.

Note that if the overload protector is an integral part of the defective relay which has been removed, there will be no overload protection for the motor. A suitable overload must be selected and fitted.

Figure 39 shows the electrical connections for an IC-22 relay.

Capacitors

Those normally used on smaller units are of the electrolytic type. They can be considered as being an electrochemical component employed to improve the phase angle relationship between the motor windings when the motor starts and runs.

They may be installed in a series or parallel with the motor windings. If a specific capacitance is not available as a direct replacement, two or more capacitors may be used. The capacitance is stamped on to the casing of a capacitor, and its value is given in microfarads.

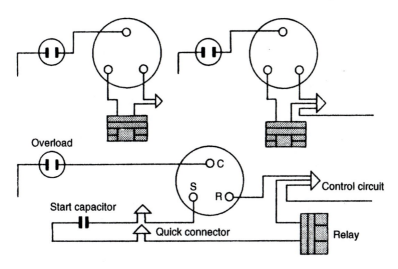

Overload

Start capacitor

Quick connector

Control circuit

Relay

Figure 39 *IC-22 relay circuits*

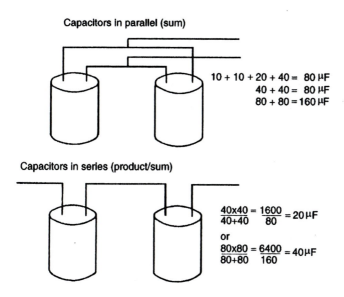

Capacitors in parallel (sum)

$10 + 10 + 20 + 40 = 80\ \mu F$
$40 + 40 = 80\ \mu F$
$80 + 80 = 160\ \mu F$

Capacitors in series (product/sum)

$\frac{40 \times 40}{40 + 40} = \frac{1600}{80} = 20\ \mu F$

or

$\frac{80 \times 80}{80 + 80} = \frac{6400}{160} = 40\ \mu F$

Figure 40 *Capacitor arrangement*

A method of selecting and connecting capacitors for making up specific capacitances is given in Figure 40.

Centrifugal switch (open-type motors)

This form of starting device is used on open drive motors. It comprises three major components: the switch contacts, the moving contact arm and the governor assembly (see Figure 41).

The switch contacts are wired in series with the supply and the start winding.

The governor assembly is mounted on the motor shaft. It consists of a spring loaded weight on a small rod. The spring loaded contact arm has a circular inclined plane extending around the shaft; this is commonly known as the skillet plate.

When the motor is idle, the switch contacts are closed. When the motor starts, the shaft rotates, gathers speed and sets up a strong centrifugal force which causes the weight to move outwards against the spring tension. This

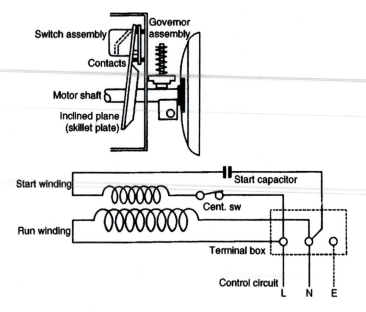

Figure 41 *Centrifugal switch*

permits the contact arm to move and open the switch contacts, thus discon-
necting the start circuit.

As the supply is disconnected to the run winding by the action of the
temperature control, the motor speed reduces and the centrifugal force
weakens. This allows the governor to move in towards the shaft, contacting
and depressing the skillet plate and closing the switch contacts ready for a
restart.

Motor protection

The most common cause of motor failure is overheating. This condition is
created when a motor exceeds its normal operating current flow. The result
can be either a breakdown of the motor winding insulation and a short circuit,
or a winding burn-out. For this reason, overload protection is provided in the
form of a current and temperature sensitive control which will open the circuit
before any damage to the motor can occur.

The following types of control are used: fuses, circuit breakers, bimetal
switches and thermistors. Fuses and circuit breakers are located remotely and
are normally required to protect the circuit conductors.

Bimetal switches are accepted as an overload protection for most hermetic
and semi-hermetic motor compressors (Figures 42 and 43).

A thermistor is a solid state semiconductor which heats up as current is
passed through it. As the temperature of the material of which it is made

Contacts closed   Contacts open
Normal running   Overload condition

Figure 42 *Typical external overload protector*

Figure 43 *Typical overload fitted into motor windings*

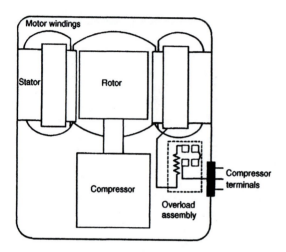

Figure 44   *Typical internal overload protector*

increases, a greater resistance to current flow is created and under overload conditions the current flow very quickly almost ceases, thereby stopping the compressor. Most thermistors are made of lithium chloride or coated barium titanate. The resistance changes approximately 6 per cent of each degree C.

Thermistors that have a PTC are connected in series with the windings and prevent current flow when the temperature of the motor increases.

The negative temperature coefficient (NTC) device is a small module which is embedded in the motor windings. Internal overload protectors perform the same function as the external and thermistor types (Figure 44).

Motor overheating has various causes. Motor compressors used for refrigeration duty are designed to start when there is an equalizing of pressures within the compressor. An increased starting load such as a high head pressure can cause overheating and excess current to be drawn. A shortage of refrigerant can also result in overheating of the motor windings, since the motors rely on returning suction vapour to assist in the cooling of the windings.

Electrical test procedures

Supply and control switch

1   Check the electrical supply at the power point, the fuse and the fuse rating.
2   Ensure that the control switch is set to the on position, and check the electrical supply to the control. If there is supply to the input side of the

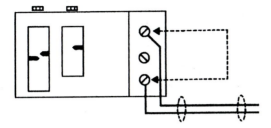

Figure 45 *Input circuit of control test*

control, bridge out the control contacts with a jumper wire or join the two control wires together; this will effectively take the control out of circuit (Figure 45).
3 If the compressor fails to start, check the supply voltage at the compressor terminals or at the relay terminal board.

Overload protector

If the compressor is cold to the touch it is evident that it has been idle for some time. The overload protector should be checked for continuity or by bridging the two contacts as described for the control switch (Figure 46). Try to start the unit.

Windings

Isolate the unit electrically and remove the electrical leads from the compressor terminals, or the relay if it is a push-on type.

Check the compressor windings for continuity and resistance values (Figure 47). The highest resistance value between any two terminals is the

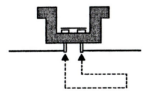

Figure 46 *Overload protector test*

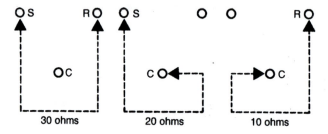

Figure 47 *Compressor motor winding resistance*

sum of two winding resistances (start and run terminals). The next highest resistance value between two terminals is the value of the start winding (start and common terminals). The lowest resistance value is that of the run winding (run and common terminals).

Test cord

If the check shows that the continuity and resistances are satisfactory, connect a test cord (see Figure 48). With the test cord connected, switch on the power supply. At this stage the compressor will be trying to start on the run winding only and a humming sound should be audible.

Depress the test cord bias switch to bring the start winding in circuit, and hold for 2–3 seconds. The compressor should start. If the compressor starts when the start winding is in circuit but stops when the bias switch is released, the run winding is defective. Failure to start at all means that the start winding is defective. In either case a replacement compressor will be required.

A maximum deflection recorded on the ohmmeter indicates that the compressor motor is down to earth, and will probably blow fuses when the windings are energized.

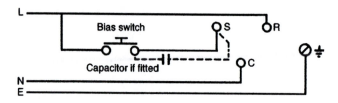

Figure 48 *Typical test cord arrangement*

When capacitors are included in the circuit they should be connected in series if a test cord is used; Figure 48 shows a start capacitor in series with the start winding.

The larger hermetic and semi-hermetic 220/240 volt a.c. 50 hertz capacitor-start/capacitor-run motor compressors draw considerable starting currents and it is advisable to use a heavy duty test cord similar to the type shown in Figure 49 when testing a compressor operation. The connection procedure is as follows:

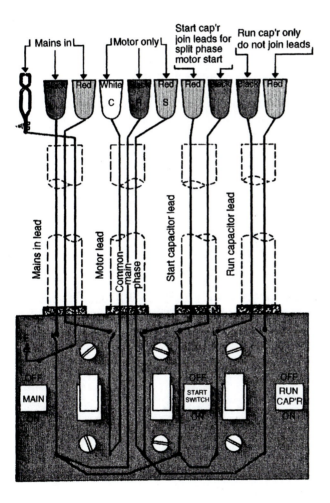

Figure 49 *Heavy duty test cord*

1 With all switches in the 'off' position, attach the three motor leads of the test cord to the compressor terminals.
2 Attach a known good capacitor of the correct microfarad rating to the 'start capacitor' leads of the test cord. (If two start capacitors are fitted, connect them in parallel.)
3 Attach the appropriate capacitor to the 'run capacitor' leads of the test cord.
4 Move the 'run capacitor' switch to the 'on' position.
5 Ensure that the earth lead of the test cord locates to the compressor screw and makes good contact.
6 Connect the test cord to the mains supply.
7 Hold the start switch at the 'on' position and move the main switch to the 'on' position.
8 Release the start switch when the motor is up to speed.

If the compressor fails to operate it must be replaced. If the compressor runs on the test cord, then the fault may be in the overload protector or starting relay.

Capacitors

If the unit fails to start with the test cord it could also mean that the capacitor is unserviceable. It should be tested with a capacitor tester if available.

Another means of testing a capacitor is to connect it across a 50 Hz supply in series with an ammeter and several lamps to provide a resistive load. The voltmeter should be connected across the capacitor terminals as shown in Figure 50.

Capacitance in microfarads (μF) is given by (amperes/volts) × 3200. For example, suppose a capacitor rating is 80–105. If the mains supply is 240 V 50 Hz and the current drawn is 7.5 A, then $(7.5/240) \times 3200 = 100$ μF. This shows that the capacitor is functioning within the design rating. It must be emphasized that this method must not be used to select a capacitor.

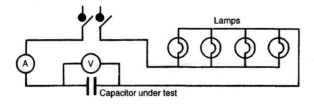

Figure 50 *Capacitor testing*

Should a capacitor show any signs of leakage or damage to the outer casing, it should be changed.

Potential relay

First ensure that no capacitor in the circuit is defective or shows signs of leakage or damage.

1 Check for continuity by removing the relay and measuring the resistance across the relay coil (see Figure 51). In this case the reading is taken between terminals 2 and 5, but this may differ according to the relay type. A high resistance should be recorded. If a zero reading is recorded then the coil is open circuit.
2 Replace the relay and switch on the unit. If the contacts are stuck open, a humming sound should be heard; the compressor is trying to start on the run winding only. After 15–20 seconds the motor will trip out on the overload.
3 Isolate the unit and install a jumper wire across the switch terminals, in this case between terminals 1 and 2. Switch on the unit. The jumper wire should include a bias switch for safety.
4 Hold the bias switch in circuit for approximately 5 seconds to allow the compressor to reach its design speed. Then release the bias switch; the compressor should continue to run. This means that the relay contacts are stuck open.

This test will only be effective if the capacitors are serviceable, and is similar to using a test cord.

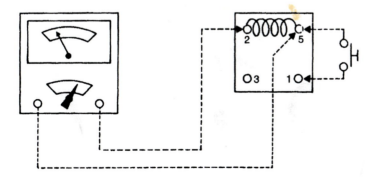

Figure 51 *Testing a four-terminal potential relay*

8 Domestic refrigerators and freezers

Because of the numerous designs of these appliances, a description of the system pipework layout is helpful to the service engineer.

In comparison with commercial or industrial installations, domestic systems are more prone to motor failure and subsequent contamination because few users appreciate the need for adequate air circulation. The method of charging these systems is different because of the minute operating refrigerant charge involved, and the decontamination and evacuation procedure also differs to some extent.

In modern production methods, domestic appliances are assembled and insulated *in situ*, with an expanded foam insulant. The foam insulant sets and bonds itself to the components it is intended to insulate. The welding technique of fusing aluminium to components and system tubing is often the source of refrigerant leakage. To gain access to the leak area is uneconomical and sometimes not possible without damaging the appliance structure. Overall, therefore, servicing may be somewhat limited.

Many service and repair outlets will not undertake repairs to an evaporator which has developed a leak. A large amount of moisture could have been drawn into the system after the leak had developed by the continuous running of the compressor. This will contaminate the compressor motor, which could fail after a comparatively short operating period following the repair or replacement of the evaporator.

Whenever a system has been subject to moisture contamination it must be thoroughly evacuated. It is often advisable to do this under workshop conditions rather than on customer premises, and this again raises the question of economical viability.

Appliance systems

There are numerous refrigerator and freezer system arrangements. Figures 52–55 show the basic refrigeration circuits for a conventional

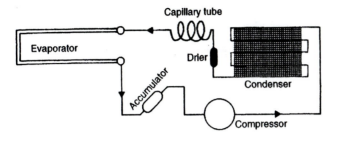

Figure 52 *Conventional refrigerator*

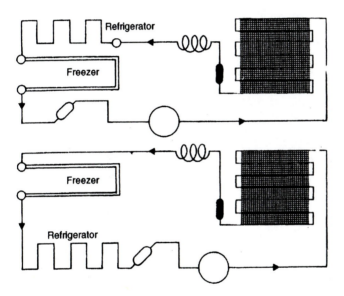

Figure 53 *Two-door refrigerator/freezer*

refrigerator, a two-door refrigerator/freezer, an upright freezer and a chest freezer.

Components and operations

Capillary restrictor

Domestic refrigerators and freezers do not employ a mechanical device as a refrigerant flow control. Instead a capillary restrictor meters refrigerant liquid

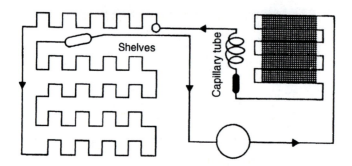

Figure 54 *Vertical freezer*

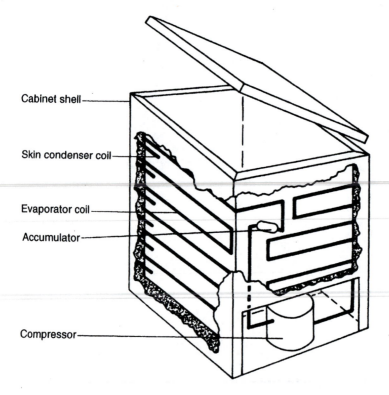

Figure 55 *Chest freezer*

to the evaporator and maintains a pressure differential whilst the compressor is operating.

Basically the capillary restrictor is a small tube. The refrigerant flow rate is determined by the length of the tubing and the internal diameter of the bore. Refrigerant will continue to flow through the capillary when the compressor stops until the pressures in the system (high side and low side) equalize.

The capillary is normally located after the filter drier; it is sometimes formed into a tight coil around the suction line. Figure 56 shows a typical domestic refrigerator arrangement. It will be noted that the capillary actually passes through the inside of the suction line to the evaporator and thereby provides a heat exchange feature which improves compressor performance.

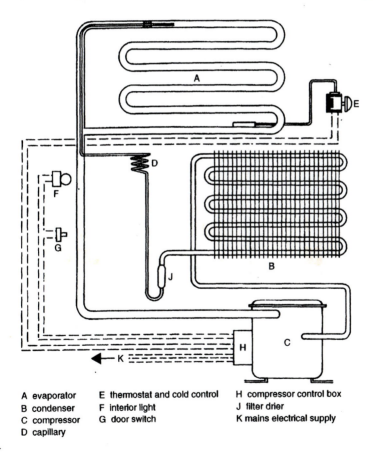

A evaporator	E thermostat and cold control	H compressor control box
B condenser	F interior light	J filter drier
C compressor	G door switch	K mains electrical supply
D capillary		

Figure 56 *Domestic refrigerator details*

Accumulator

Suction line accumulators are employed to prevent frosting back along the suction line after off cycles. This is because the relatively small refrigerant charges in modern refrigerators and freezers are difficult to control accurately, and some overspill from the evaporator occurs when the compressor stops.

Accumulators are shown in Figures 52–55.

Freezer door heater and oil cooler

Figure 57 shows a freezer system diagram with door frame heater and oil pre-cooler.

Hot gas defrosting

Figure 58 shows the normal refrigeration and defrost cycles in an appliance using hot gas defrosting.

Solenoid flow control for refrigerator/freezer

Figure 59 shows the refrigeration system and Figure 60 the electrical circuit for solenoid flow control.

When the refrigerator thermostat is made and the freezer thermostat is open circuit, the relay is energized through contacts 1 and 3. The compressor operates but the solenoid is not energized, and the refrigerant flows through the refrigerator evaporator and the freezer evaporator.

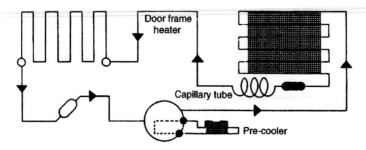

Figure 57   *Freezer door frame heating and oil cooling*

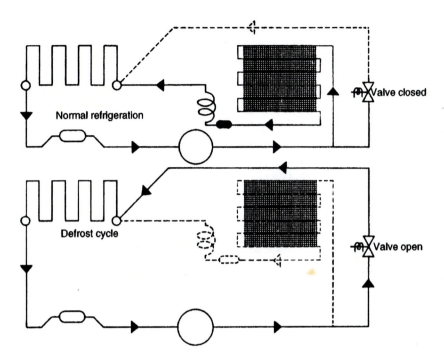

Figure 58 *Hot gas defrosting*

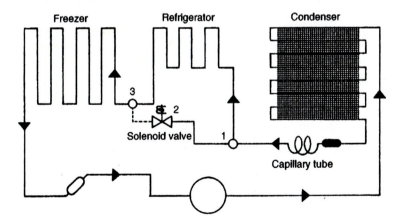

Figure 59 *Refrigerator/freezer with solenoid flow control*

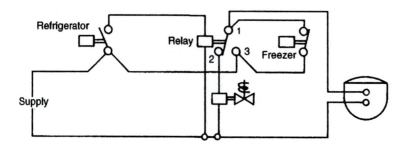

Figure 60   *Flow control circuit*

When the freezer thermostat is made and the refrigerator thermostat is open circuit, the compressor operates and the solenoid is energized through contacts 1 and 2. The refrigerator evaporator is by-passed.

When the refrigerator thermostat makes whilst the compressor is operating, the relay is energized. This opens contacts 1 and 2, and refrigerant will again flow to the refrigerator evaporator as well as the freezer evaporator.

Electrical faults

The diagnosis of electrical faults on components and compressors has already been dealt with in Chapter 7. The following procedure may be carried out to quickly pinpoint the faulty component and to eliminate unnecessary dismantling. The electrical circuit is tested for continuity from the terminal board of the circuit. The typical electrical circuit in Figure 61 has been chosen for the test procedure.

Steps 1 and 2 of the test procedure can be disregarded if the refrigerator or freezer has an interior light. This will come on when the appliance door is opened, unless the light bulb itself is defective.

The procedure is as follows:

1   Test for continuity from points 1 to 2, which will test the whole circuit.
2   Test from point 3 to points 1 and 2 to reveal any earth fault.
3   Test from points 1 to 8, which eliminates half of the circuit.
4   If the fault is in section 1 to 8, test between points 1 and 5 to determine if the fault is due to a loose connection, a defective cable or a blown fuse.
5   Should the fault be located in section 5 to 8, then the thermostat must be considered inoperative because the interior light and door switch will be isolated when the door is closed.

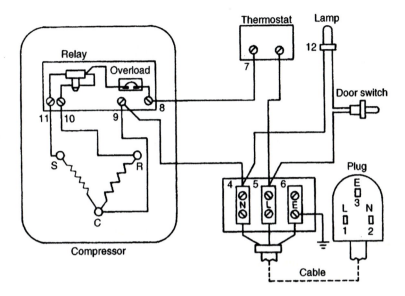

Figure 61 *Domestic appliance circuit testing*

6 If the fault is in the compressor half of the circuit, it should be located in a similar manner by dividing into sections until it has been finally determined.

When an ohmmeter is used to test for continuity, it should normally be set to read off the ×1 scale. If the total resistance of the circuit is in excess of 100 ohms, the ×10 scale will be required; reset to the ×1 scale when the faulty section is isolated to less than 100 ohms.

Figure 62 shows a typical refrigerator/freezer electrical circuit.

Decontaminating domestic systems

With liquid flushing using R11 no longer permissible the previously described dilution method can be carried out (see Chapter 16).

Refrigerant charging in domestic appliances

The principles of refrigerant charging have been covered in Chapter 4.

Domestic refrigerators and freezers operate with very small refrigerant charges, and the charge must be administered accurately. This can be best

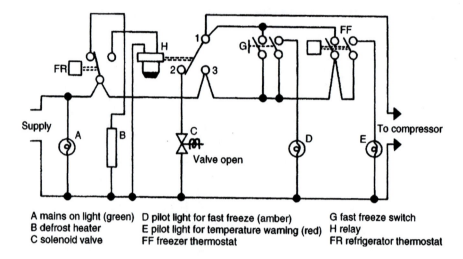

A mains on light (green)   D pilot light for fast freeze (amber)   G fast freeze switch
B defrost heater   E pilot light for temperature warning (red)   H relay
C solenoid valve   FF freezer thermostat   FR refrigerator thermostat

Figure 62   *Typical refrigerator/freezer electrical circuit*

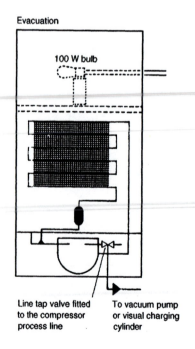

Line tap valve fitted   To vacuum pump
to the compressor   or visual charging
process line   cylinder

Figure 63   *Decontaminating domestic appliances*

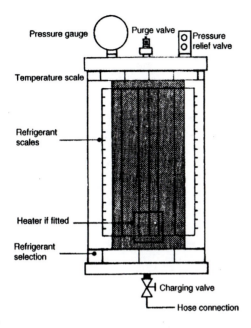

Figure 64 *Visual charging cylinder*

achieved by using a visual charging cylinder, sometimes called a dial-a-charge (Figure 64).

This is basically a small refrigerant cylinder with a liquid-indicating sight glass or tube. In front of the sight glass and surrounding the cylinder and sight glass is a rotating screen upon which are graduated scales for the common refrigerants, R12, R22 and R502. Above the refrigerant scales is a temperature range.

Fitted to the top of the cylinder assembly are a pressure gauge; a purging valve, which is normally a Schraeder type; and, if a heater is incorporated, a pressure relief valve for the cylinder protection. At the base of the cylinder is a charging valve and hose connection.

Filling the cylinder

1 Obtain a service cylinder of the correct refrigerant. Connect it to the visual charger with a suitable hose.
2 Open the valve on the service cylinder and invert the cylinder.

3   Slacken the hose connection at the charging valve to purge air from the hose. Tighten the hose connection.
4   Open the charging valve to allow refrigerant liquid to flow by gravity into the cylinder. Observe the sight glass: when liquid flow ceases, depress the purge valve and gently vent off a small amount of refrigerant vapour. When a pressure difference between the charging cylinder and the service cylinder is created, the liquid will begin to flow again. Repeat this operation until the required amount of refrigerant is in charging cylinder.
5   Close the valves on both the visual charger and the service cylinder.
6   Disconnect the service cylinder.

Charging the system

1   Assuming a line tap valve has been installed and the system evacuated, connect the hose to the charging valve and the line tap valve.
2   Open the charging valve and slacken off the hose connection to purge air from the hose. Close the valve.
3   Note the ambient temperature. Rotate the screen until the required refrigerant scale lines up with the ambient temperature on the temperature scale.
4   Note the level of the refrigerant in the sight glass.
5   Open the line tap valve fully and slowly meter the prescribed amount of refrigerant into the system. If the visual charger has a heater this can be energized before opening the line tap valve to create a pressure difference between the charging cylinder and the system.
6   When the charge has been administered as indicated by the sight glass, close the charging valve.
7   Allow a few minutes for the liquid in the hose to vaporize and the system pressure to equalize. Then close the line tap valve.
8   Disconnect the visual charger and leak test the system.

Alternative method of charging

If a refrigerant charge cannot be accurately measured by using a visual charger, it must be drawn into the system by the compressor from the low side. It is imperative that this is carried out slowly to eliminate the risk of overcharging, which could damage the compressor.

By allowing small amounts of refrigerant vapour into the system, and observing the frost line on the evaporator, overcharging can be prevented.

When the frost line reaches the location of the thermostat bulb it is always advisable to stop charging and allow sufficient time for the system to reach average evaporating temperature and start to cycle. Should frosting back occur, purge off the surplus refrigerant in small amounts from the line tap valve connection.

Reference to the compressor nameplate should be made to determine the running current.

Refrigerant charges are normally stamped on the refrigerator model plate. The charge may be given in ounces or grams according to the age of the refrigerator. These quantities can be easily converted: 1 ounce = 28.35 grams. For example, 150 grams is equivalent to 150/28.35 = 5.29 ounces; 6.5 ounces is equivalent to 6.5 × 28.35 = 184.27 grams.

Remember, an undercharge of refrigerant will result in long running periods or continuous running. An overcharge of refrigerant will result in frosting back, overheating, increased running costs and could possibly damage the compressor.

The domestic absorption system

The modern domestic absorption refrigerator or 'gas refrigerator' operates on the principle of using heat to produce cold. It employs a mixture of hydrogen and ammonia with water (aqua-ammonia) as the cooling agent: water has an affinity for ammonia, and the hydrogen speeds up the process of evaporation.

The heat source may be town gas or, in the case of smaller systems designed for use in caravans, camp sites etc., propane, butane or even paraffin. An all-electric version has a small heater element to provide heating.

The valve assemblies which regulate the gas flow incorporate a small pilot jet to ignite the gas, as the valve is energized electrically. The evaporating temperature is controlled by a thermostat which in turn energizes the gas valve or heater element, depending upon the setting of the thermostat (cold control).

Operation

When the thermostat energizes the valve or heater, heat is applied to the ammonia and water mixture in the generator. The liquid then boils off the pass through a small tube (percolator tube), in much the same way as coffee in a percolator, to enter the separator or rectifier. From there it circulates by gravity.

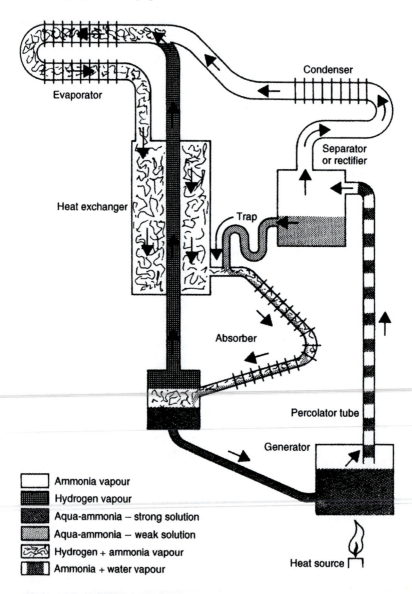

Figure 65 *Absorption system cycle*

The ammonia vaporizes faster than the water and the aqua-ammonia separate in the rectifier; the ammonia vapour rises to the condenser and the water drains back to the absorber to be recirculated. The lighter ammonia vapour rising from the condenser coils changes back to a liquid to drain into the evaporator, where it mixes with the hydrogen vapour from the absorber.

Heat from the stored product in the food compartment of the refrigerator will vaporize the ammonia liquid in the evaporator. The mixture of ammonia vapour and hydrogen, being heavier than either of the two gases alone, passes to the absorber where it meets the water coming from the rectifier. The ammonia is then absorbed by the water and, when the water has absorbed as much vapour as it can hold, the vapour returns to the evaporator.

At maximum working conditions (maximum heat at generator) the pressure on the high side of the system (condenser and absorber) will be approximately 14 bar with a relative temperature of 36 °C. Assuming that the refrigerator is in an average ambient temperature of 21 °C, the heat rejection will be quite reasonable with a 15 ° temperature (36−21 °C). Pressure on the low side of the system will be 12.6 bar hydrogen and 1.2 bar ammonia, so the relative temperature of the evaporator will be −15 °C. At this temperature, heat exchange from the food products to the evaporator will be at an average 3 °C, producing a 12 ° temperature difference [3 to − 15 = 12 °].

Servicing

Since this is a completely sealed system containing two gases, very little can be achieved by the service engineer except the replacement of a thermostat, valve assembly or heater element.

The pilot jet and valve should be kept clean and the refrigerator should be carefully levelled at the time of installation and checked when service is carried out.

Figure 65 shows the absorption system cycle.

9 Fault finding guide for vapour compression systems

This checklist will assist in the quick diagnosis of faults on vapour compression systems.

Visual fault finding

Compressor not running

1 Main supply isolator open.
2 Fuse blown.
3 Overload open circuit.
4 Control circuit open (low pressure switch, high pressure switch or thermostat).

Fixture temperature too high

1 High evaporator load.
2 Defrost heater(s) energized.
3 Condenser blocked with dirt.
4 Restricted air flow over the condenser or restricted water flow.
5 Condenser fan inoperative.
6 Evaporator blocked with ice.
7 Evaporator fan inoperative.
8 Vapour bubbles in sight glass (possible shortage of refrigerant).
9 Frosting liquid line (blocked drier).
10 Broken or loose drive belts.
11 Expansion valve bulb loose on suction line or broken capillary.

System noisy

1 Evaporator/condenser fan blades touching fan guards or loose on the motor shaft.
2 Compressor oil sight glass empty (shortage of oil or oil entrainment).
3 Compressor head frosted (compressor pumping liquid refrigerant).
4 Loose motor pulley, compressor flywheel, compressor mountings.

Pressures

If the fault is not obvious after visual inspection:

1 Attach compound and pressure gauges.
2 Check that the compressor is pumping.
3 Observe operating pressures.

High discharge pressure

1 Overcharge of refrigerant.
2 Air in the system.
3 Dirty condenser, poor air supply.
4 Inadequate water flow.
5 High load imposed on the evaporator.

Low discharge pressure

1 Shortage of refrigerant.
2 Compressor inefficient.

High suction pressure

1 Compressor inefficient.
2 Overcharge of refrigerant (capillary systems only).
3 Expansion valve defective.
4 High evaporator load.
5 Defrost system operating when compressor is running.

Low suction pressure

1 Shortage of refrigerant.
2 Blocked or defective expansion valve.
3 Blockage in liquid line, drier solenoid valve or shut-off valve.
4 Blocked evaporator (excessive ice build-up).
5 Inoperative evaporator fan.
6 Defective water pump or blocked water filter on a chiller system.

Note: a temperature difference across any component in the liquid line is an indication of a partial restriction.

Advanced diagnosis

Expansion valve capacity too small

1 No subcooling of the refrigerant liquid.
2 The pressure drop across the expansion valve is less than that for which it was dimensioned.
3 Incorrect expansion valve bulb location (too cold or being cooled).
4 Large pressure drop across the evaporator.
5 Expansion valve blocked by ice or foreign objects.
6 Incorrect expansion valve external equalizer location.

Liquid knock in compressor during start-up

1 Compressor discharge valve letting by (refrigerant condenses in the discharge line to enter the compressor).
2 Low ambient temperature in condensing unit location (refrigerant condenses in the compressor).

When refrigerant vapour condenses in the discharge line, which is common in equipment located outside and during winter operation, or with low ambient temperature combined with long 'off' cycles, liquid refrigerant can enter the cylinders if discharge valves are not seating correctly. This condition can be attributed to poorly designed pipework (see 'Pipework and oil traps' in Chapter 10).

3 Suction line passing through low ambient temperature, or not insulated when installed in a low temperature coldroom (refrigerant condenses in suction line).

When condensation occurs in the suction line, refrigerant will be drawn into the cylinders when the compressor starts.

4   Suction line has a free fall to the compressor.

This result of poor pipework design can be overcome by installing an oil trap in the suction line.

5   Expansion valve bulb mounting gives poor thermal contact, or the bulb is located in a warm place.

The expansion valve thermal element will respond to temperature. If it is incorrectly located it can produce overfilling of the evaporator during operation, which may allow liquid refrigerant to enter the suction line and compressor. This is more likely to occur with short suction line runs.

Liquid knock in the compressor during operation

Evaporator pressure too low, or suction superheat too narrow

Suction pressure too low, or suction superheat too wide

A low pressure reading on the compound gauge during operation does not necessarily indicate malfunction. The pressure reading must be related to the design temperature and the temperature difference recommended to the product being stored.

Single phase/three phase motor compressors and remote-drive motors

Compressor will not run

1   Fuse blown.
2   Main isolator open.
3   No supply at motor terminals.
4   Overload open circuit.
5   Control circuit open.
6   Burnt-out motor windings.

Compressor hums but does not start

1   Incorrect wiring to motor winding (3 ph).
2   Low line voltage.

3 Start capacitor open circuit.
4 Relay inoperative.
5 Motor winding open circuit.
6 Seized compressor.
7 Piston or impellor jammed by broken component.

Compressor does not attain design speed

1 Incorrectly wired motor winding (3 ph).
2 Low line voltage (3 ph).
3 Defective relay.
4 Start capacitor defective.
5 High discharge pressure.
6 Star/delta motor compressor not unloaded (3 ph).

Compressor short cycles

1 Control differential too narrow.
2 Valve plate letting by.
3 Motor overloading.
4 Shortage of refrigerant.
5 Expansion valve defective.
6 Restriction in system.
7 High pressure cut-out set too low (switch operating too soon).

Part Two
Installation and Commissioning

10 *Pipework and oil traps*

Installation principles

The design and correct installation of the system pipework contributes to the reliability of the equipment, irrespective of the type. Design problems seldom occur with packaged systems because they would be overcome during development, but attention must be paid to other systems.

When a large installation is programmed, a site survey is usually carried out. This is intended to plan pipework routes, the method of installing and the components necessary to ensure a trouble-free installation, and to foresee any snags which might lead to costly modifications afterwards. In many instances the installation of small systems is left to the discretion of the installing engineer.

A reliable system must take into account the following:

1 The condensing unit should be mounted on a level and solid foundation.
2 Accessibility to all components for service and maintenance should be ensured.
3 Tubing runs between the evaporator and the condensing unit should be made as straight as possible and by the shortest route.
4 Excessive pressure drop should be avoided.
5 Where possible, depending upon the method of fixing, suction and liquid lines should be run together to produce an effective heat exchange.
6 Oil traps must be provided to ensure adequate oil return to the compressor.
7 Pipework must be firmly supported.
8 Where pipework passes through walls, floors and ceilings suitable sleeves should be provided. When routed at floor level, pipework should be protected.
9 The structure upon which the pipework is to be installed should be examined carefully.
10 Flexible couplings should be provided where necessary.

It is not intended to cover the finer details of installation practice.

Pipework fittings

Although it is general practice to install pipework where possible with brazed joints, flare fittings are still used. These should be kept to a minimum because too many in a circuit will create a pressure drop.

Flare unions and service and shut-off valve connections do not determine the diameter of the system pipework. It may be necessary to change a flare union to eliminate a pressure drop across a component.

A common cause of failure is the partial or complete loss of the refrigerant charge resulting from the use of an incorrect flare nut. A frostproof flare nut, for example, should be used on suction lines where extreme temperature changes can take place, i.e. frost on the suction line melting during an off cycle and freezing when the unit restarts. It is possible for ice to form between the long shoulder of a standard flare nut and the tubing; expansion of the ice and contraction of the tubing eventually fractures the flare (see Figure 66).

When a frostproof type is used on the high side of the system, vibration from gas pulsations at the compressor discharge or from an evaporator fan, plus the weight of the expansion valve, can cause the flare to rupture. The long shoulder

Frostproof flare nut (short)   Standard flare nut (long)

Figure 66   *Flare nuts*

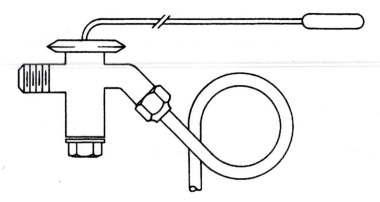

Figure 67   *Vibration loop*

1/4 in equal flare union   1/8 in Briggs gas half union

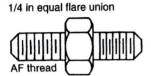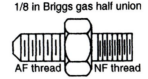

AF thread   AF thread   NF thread

Figure 68   *Flare unions*

of the standard flare nut can prevent this by virtue of the reinforcement it provides, restricting lateral movement of the tubing and flare.

A vibration loop formed in a liquid line or in the small diameter tubing of a control line can have a twofold advantage. It will protect the flare against vibration, and may allow easy reforming of the tubing when a valve or control has to be changed (see Figure 67).

The leak potential is also increased when a flare union is used instead of a half union for connecting control lines to compressors or components. A half union does not have the seat facility of a flare union, although in some instances the threads may be the same. The difference is depicted in Figure 68.

Pipework supports

Currently three types of supports are used: the tubing saddle for smaller diameter tubing; a more elaborate assembly such as Hydra-zorb for larger diameter tubing; and the Munsen ring for iron and steel pipework, which obviously requires a stronger type of support structure (Figure 69).

Pipework must be well supported. For soft drawn copper tubing, the number of saddles used will depend upon the length of tubing run. For long runs the suction and liquid lines should be supported at intervals not exceeding 30 suction line tube diameters.

Hydra-zorb is ideal for larger hard drawn copper tubing and multiple installations. It consists of a metal wall mounted channel into which resilient tube mounting blocks and clamps are inserted.

The Munsen ring is basically a split clamp, which is usually cemented into the wall.

Soft drawn copper tubing supported by saddles brings both suction and liquid lines into close proximity. The tubing can be easily formed to touch and make thermal contact, thus creating heat exchange. The effect is to subcool the liquid refrigerant passing to the expansion valve and superheat the vapour returning to the compressor, which improves the system efficiency.

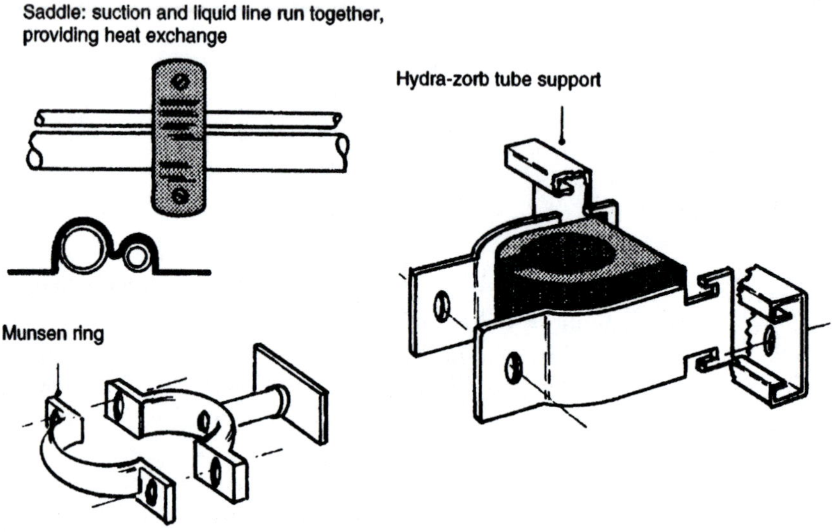

Saddle: suction and liquid line run together,
providing heat exchange

Hydra-zorb tube support

Munsen ring

Figure 69 *Pipework supports*

The other two types of support do not allow this heat exchange. Therefore for those types the tubing should be insulated where there is a possibility of frost or condensation forming.

Isolating valves and controls should be installed in the pipework so as to provide easy access for maintenance or replacement.

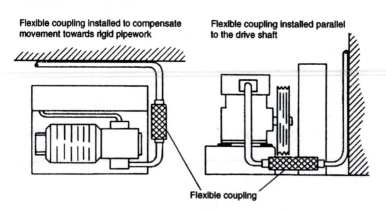

Flexible coupling installed to compensate
movement towards rigid pipework

Flexible coupling installed parallel
to the drive shaft

Flexible coupling

Figure 70 *Flexible coupling locations*

Pipework should not obstruct the view of the crankcase oil sight glass. It should also be routed to avoid motor compressor end plates and terminal covers, and the drive belts and motors of open-type units.

When compressors are resiliently mounted it is advisable to install a flexible coupling (vibration eliminator) between the compressor and the pipework to compensate for movement when the compressor stops or starts. This is common with hermetic and semi-hermetic motor compressors. With open-type units this may not be necessary, but if a flexible coupling is used it should be installed parallel to the compressor drive shaft. These couplings are illustrated in Figure 70.

Pipework routes

When condensing units or remote condensers are located above the level of a coldroom, it is possible that the pipework will pass through a wooden floor. The pipework will be subject to variations in temperature. Condensation can form on a suction line, and if this is insulated the insulation can become saturated.

This will encourage wet rot and fungal growth. In addition, the combination of a warm liquid line and wet materials will create an ideal environment for bacteria.

To overcome this problem a sleeve of metal or plastic, preferably the latter, should be inserted into the pipework aperture (Figure 71). The pipework is routed in the sleeve, which can then be sealed with a moisture and fire resistant compound. The sleeve is anchored firmly to a joist to avoid movement. Effective sealing of the sleeve prevents ingress of vermin. It also blocks off a supply of air, which would be undesirable if a fire started in the space below the floor. The sealing compound, being resilient, will prevent damage to pipework due to vibration.

The same principles should apply when pipework passes through a brick wall. The sleeve provided for an outside wall must protrude 25 mm to deny entry of rain water into the sleeve (see Figure 72).

Under no circumstances must electrical cables occupy the same sleeve as the refrigerant carrying pipework.

A condensing unit may be located on the ceiling of a small coldroom. It must be mounted on load bearing members; the weight of the unit will then be taken up by the supporting walls of the coldroom, thus preventing any sagging of the ceiling and subsequent breakdown of the insulation. An example is given in Figure 73.

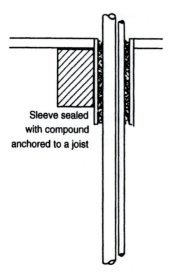

Sleeve sealed
with compound
anchored to a joist

Figure 71 *Pipework passing through flooring*

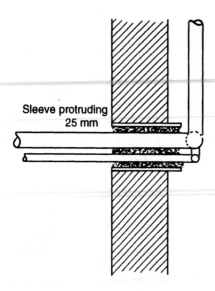

Sleeve protruding
25 mm

Figure 72 *Pipework passing through wall*

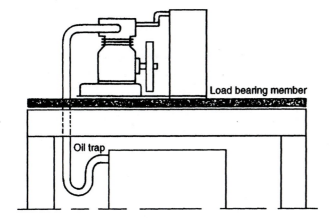

Figure 73 *Ceiling mounted condensing unit*

Oil traps

Suction line oil traps

High suction line risers will inevitably lead to compressor lubrication problems if provision for oil return is not made. When the compressor is located above the evaporator, an oil trap must be installed.

During low suction pressure operation at the end of a cooling cycle, the oil will tend to separate from the suction vapour and cling to the surface of the tubing. It will drain back to become entrained in the evaporator during off cycles. The oil trap will collect the oil, and when the compressor restarts the higher suction pressure will return most of it to the compressor.

An excessive amount of oil in the evaporator will reduce the evaporator capacity.

Oil traps can be of a U formation or of the barrel type; both perform the same function. A typical suction oil trap, formed in the suction line during installation, is depicted in Figure 73. On larger installations the higher suction riser will need more oil traps, as depicted in Figure 74.

Discharge line oil traps

These are installed or formed in high discharge risers (Figure 74) to prevent oil from draining back to the compressor head during off cycles. If it is present in the head a dynamic pressure will develop as the compressor restarts,

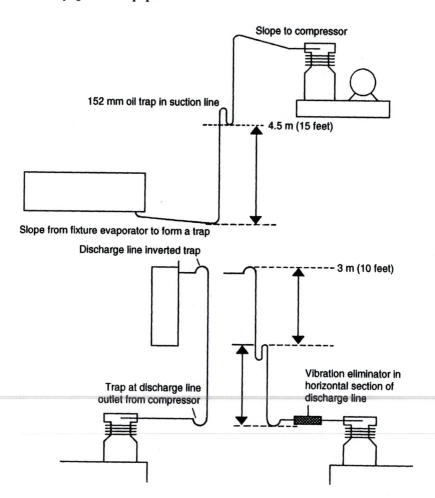

Figure 74   *Oil traps*

resulting in damage to the compressor valves, blown gaskets and possibly broken connecting rods and drive shafts.

Oil separators

A certain amount of oil leaves a compressor with the discharged refrigerant vapour. Large quantities may be prevented from circulating in the system by using an oil separator. Most air conditioning systems employ one of three basic types; the most common is the float type.

Oil returns to the compressor after it has been collected in the separator. The oil separates from the discharged vapour because the vapour flow slows as it enters the separator. When a certain level of oil is reached, the float opens a valve to return the oil to the compressor crankcase.

The oil return line to the compressor crankcase must be of small diameter tubing (6 mm, 0.25 in) or a capillary. This will reduce the oil flow and prevent oil slugging in the compressor.

Some separators are serviceable (bolted construction), some are installed by brazing, and the smaller types have flare connections. An oil separator does not eliminate the necessity for oil traps. Figure 75 shows a typical method of installation.

The accumulator type of separator is used in the larger installations and is located in the suction lines. The bucket type, of much larger construction, is mainly used on industrial systems.

Refrigerant may condense to liquid and collect in the oil separator during long off cycles or during a manual shut-down period of long duration. This refrigerant liquid returning via the oil return line to the compressor could damage the compressor after restart. A check valve in the vapour outlet of the oil separator will reduce the possibility of compressor damage.

A solenoid valve installed in the oil return line and interlocked with the compressor motor starter is common practice. This ensures a positive shut-off in the return line, because the solenoid valve will be energized (open)

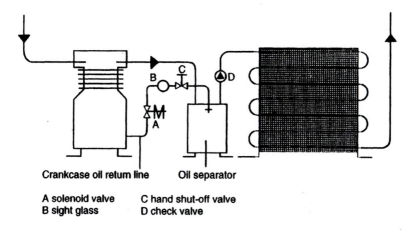

Crankcase oil return line   Oil separator

A solenoid valve   C hand shut-off valve
B sight glass   D check valve

Figure 75 *Oil separator*

only when the compressor is operating, preventing oil and refrigerant from draining to the compressor crankcase during off cycles.

To minimize refrigerant condensation in the oil separator it should be installed as close to the compressor as possible, preferably in a warm location. The separator may be insulated to retard heat loss when the compressor stops.

An indicating sight glass and a hand shut-off valve are installed in the oil return line. In conjunction with the sight glass the shut-off valve can be used as a throttling device; it is adjusted so that liquid flow (oil and refrigerant) from the separator can be controlled to the compressor crankcase.

Discharge line mufflers

In the event of noise or vibration from gas pulsations, discharge mufflers are recommended and must be installed for the free draining of oil.

As previously stated, when compressors are idle the oil adhering to the inner surfaces of the discharge line risers will drain to the bottom of the risers. With a 3 m riser, plus the discharge muffler, the amount of oil can be quite considerable.

It is therefore advisable to loop the discharge line towards the floor to form a trap so that oil cannot drain to the compressor head. This trap will also collect any liquid refrigerant which may condense during the off cycle. It is especially important when the discharge line is in a cooler location than the condenser/receiver. Mufflers are normally installed vertically near the condensing unit to provide efficient oil movement. Figure 76 shows a discharge muffler installed in a discharge line loop.

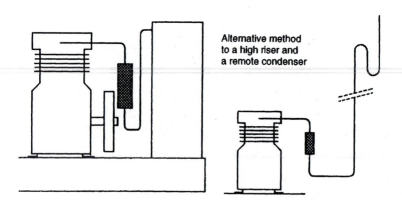

Figure 76 *Discharge mufflers*

Parallel pipework

When compressors are installed in parallel and share common suction and discharge lines, they are usually open-type units. Hermetic and semi-hermetic units are not used in case of a motor failure. The high temperatures resulting from the motor failure of one motor compressor can cause a chemical breakdown of the refrigerant and oil, resulting in the contamination of the whole system.

However, two or more hermetic or semi-hermetic units can be used to accommodate a common load. When this is required each compressor is connected to an independent circuit within the evaporator and the condenser, so that they function without the possibility of cross-contamination. A single condenser is normally preferred for ease of control, but a separate condenser may be used for each compressor.

Discharge lines

Discharge lines should be installed so that the horizontal section is pitched downwards to join the common line. This will provide a free draining trap, preventing oil drain back to the idle compressor.

When the condenser is located above the compressors, separate risers should be used. Oil traps should be included at the base of each riser if the net lift to the condenser is more than 2 m. If the riser height is less than 2 m the oil traps may be omitted.

Figure 77 shows a basic pipework arrangement for discharge lines.

Discharge line equalizing

When individual condensers are installed, discharge lines must be equalized before they enter the condensers to allow them to function as one (Figure 78).

The pressure drop within equalizing lines is critical. For example, assume a pressure drop of 0.03 bar (0.5 psig) in the equalizer line. The pressure differential between an idle and an active condenser will cause condensed refrigerant from the active condenser to back through the liquid line to the idle condenser in an effort to equalize.

Crankcase equalizing

Obviously it is also necessary to prevent unequal oil distribution between the crankcases of the compressors. This is achieved by connecting an equalizer line to tappings provided (Figure 78).

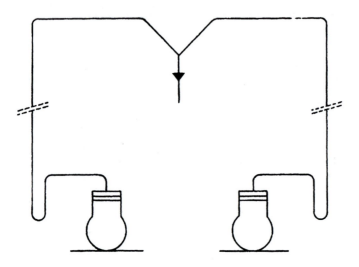

Figure 77   *Common discharge*

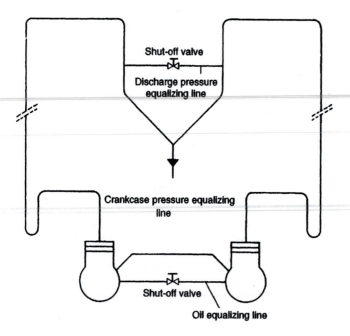

Figure 78   *Discharge and crankcase equalizing in common pipework system*

The equalizer line size must be equal to the size of the tapping union. It is also essential that the compressors are mounted on the same level or that the oil levels are at the same height; compressor mounts can be adjusted to make this possible.

Suction line equalizing

Pipework should be routed to a manifold arrangement to equalize pressure at each unit. Common suction lines will drain freely, so this method is suitable for most systems (Figure 79).

When the evaporator is at a lower level than the compressor, the suction line should be pitched downwards. There should be two oil traps approximately 0.5 m (20 in) in height when suction risers are in excess of 3 m (10 ft).

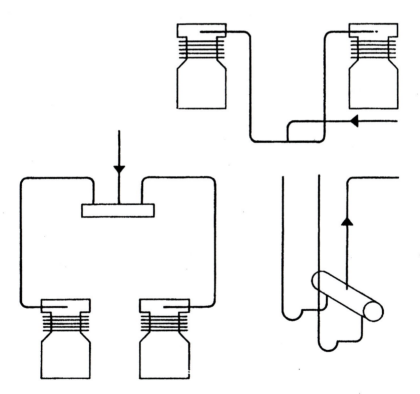

Figure 79   *Suction line equalizing*

Pipework assembly

Soft copper tubing up to 22 mm (7/8 in) diameter may be joined with flare fittings or by brazing. Hard drawn copper tubing is brazed. Iron and steel pipework is welded; cost restricts the use of welding to large industrial and ammonia systems, and it is not described here.

Flare fittings

Flare connections will be perfectly satisfactory if the flaring of the tubing is carried out correctly.

Undersized or oversized flares are potential sources of leaks. In an undersized flare the flare seat surface area is reduced and can allow movement within the flare fitting if the pipework is subjected to excessive vibration; the joint will eventually fracture. An oversized flare may prevent the correct seating to a union or a control; although it may well pass a leak test, a leak may develop when the pressures within the systems rise during operation. Flares are shown in Figure 80.

A good flare should allow withdrawal of the flare nut over the threads after it has been tightened down on to a union.

Brazing

Currently, most copper tubing is assembled with brazed joints because it is cheaper. In addition, if joints are made correctly the possibility of leaks is reduced.

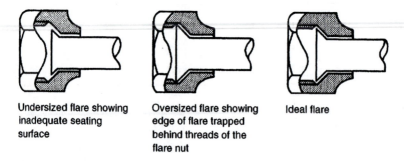

Undersized flare showing inadequate seating surface

Oversized flare showing edge of flare trapped behind threads of the flare nut

Ideal flare

Figure 80   *Copper tube flaring*

Expanding

In order to make a brazed joint the tubing has to be expanded. Various tools are available for this purpose: the punch-type expanders, swage formers used with a flare block and spinner, and the more elaborate tube expander.

The tube expander is generally used for larger diameter tubing, which may be hard or soft drawn copper. Hard drawn copper must be annealed before attempting to expand the tube. Annealing is a softening process, involving heating the tube and allowing it to cool.

When using a swage former or the expander, the depth of the swage will be determined by the former. It is approximately equivalent to the outside diameter of the tubing. Too shallow an expansion of the tube tends to produce a weak joint (see Figure 81).

Process

Brazing can be defined as jointing by applying high intensity heat to a high melting point alloy in order to fuse together two metal surfaces. It is often referred to as silver soldering.

Brazing rods should be cadmium free, and as described in BS 1845. All brazing should meet the requirements of BS 1723.

Hazards

Whilst it is possible to prefabricate some parts of the system pipework assembly in workshops, most brazing will be carried out on site.

The operator must be conversant with the site safety regulations and comply with fire regulations. Attention must be paid in particular to fire and smoke alarms whilst the brazing operations are being performed.

Where possible, work away from flammable materials such as wooden floors, joists and eaves. Use a protective metal or fire resistant sheet under conditions where a fire risk is obvious. Keep a fire extinguisher to hand when working in a risk area.

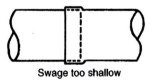
Swage too shallow

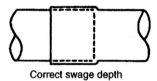
Correct swage depth

Figure 81 *Copper tube swaging*

The number of brazed joints should be kept to a minimum, and the bending of copper tubing is preferred.

Preparation

All pipework and fittings to be brazed must be cleaned to remove dirt and oxides using a fine grade steel wool. A solvent may be necessary to remove preservative coatings from components or fittings.

Heat shields should always be used to protect areas surrounding the joint and to concentrate heat. Protect the system components, which could suffer damage when heat is applied, by wrapping a damp cloth around the component close to the joint.

Ensure that all joints are a tight fit. Support the pipework before brazing is attempted; any movement before the alloy has set can result in a leak or a weak joint.

Equipment usage

Select the correct pressure and nozzle (Table 5). Always light the blowtorch with a spark gun and not with matches or a cigarette lighter. Adjust the blowtorch to give the correct type of flame (see Figure 82).

Wear protective clothing and tinted goggles or glasses. Keep oxygen and acetylene cylinders away from any heat source. Always close down the equipment valves when leaving it unattended, even for a short period. Do not smoke when brazing. Note that toxic atmospheres are created during brazing operations when cadmium, galvanized metal and paint are heated.

Table 5 *Nozzle sizes*

Thickness of tube wall		Nozzle size
mm	in	
0.9	1/32	2
1.2	3/64	4
2.0	5/64	6
2.6	7/64	10
3.2	1/8	14
4.0	5/32	20
5.0	3/16	26
6.5	1/4	36

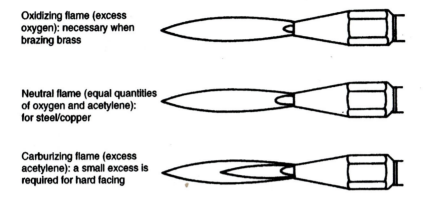

Oxidizing flame (excess oxygen): necessary when brazing brass

Neutral flame (equal quantities of oxygen and acetylene): for steel/copper

Carburizing flame (excess acetylene): a small excess is required for hard facing

Figure 82   *Brazing flame comparisons*

Always pass nitrogen through the pipework or component being brazed to prevent oxidation and the formation of scale on the interior surfaces. Clean off all joints after brazing, especially when a flux has been used.

When brazing copper tubing, the joint area should be heated broadly with a continuous circular movement of the blowtorch until the copper changes colour to a cherry red; then apply the brazing rod. This will minimize the risk of local overheating and burnt tubing. Heat should be applied indirectly so that the rod and flux are melted by conduction through the base metal.

11 *System control valves*

Crankcase pressure regulator

As a compressor starts, a heavy load is imposed on the drive motor. The crankcase pressure will be at its highest against normal or abnormal loading.

The function of the regulator is to keep the pressure in the crankcase to a reasonable level to protect the motor from overload. This applies when the evaporator pressure is above normal operating pressure, for example:

1 During the initial start-up after defrosting.
2 During periods of high starting loads.
3 Under high suction pressure caused by hot gas defrosting.
4 During surges in suction pressure.
5 Under prolonged high suction pressure.

The regulator is installed in the suction line as shown in Figure 83.

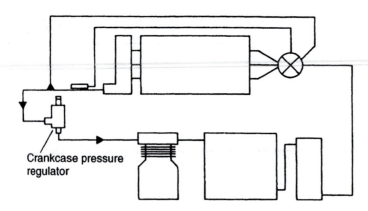

Crankcase pressure regulator

Figure 83 *Location of crankcase pressure regulator*

As a guide, the control should be set to the following crankcase pressures:

Hermetic and semi-hermetic compressors:	2.5 bar
Open-type compressors with standard motor:	3.0 bar
Open-type compressors with oversized motor:	4.5 bar.

Final adjustments are made to make the running current conform to that specified on the motor nameplate (running current rating).

Evaporator pressure regulator

This valve is fitted in the suction line to control the pressure in the evaporator, to prevent it dropping below a predetermined pressure. It is used to control the evaporating temperatures of systems such as drinking water fountains and beverage coolers, to maintain a constant evaporator pressure/temperature and prevent freezing.

It is also used in multi-evaporator systems, when it is installed in the suction line of the warmest evaporator. Check valves or one-way valves are necessary to prevent migration of refrigerant from the warmer to the cooler evaporators during off cycles (see Figure 84).

Water regulating valve

The pressure operated type is the most popular. It is employed in water cooled systems to control the flow rate through the condenser, modulating in response to changes in the condensing or operating head pressure.

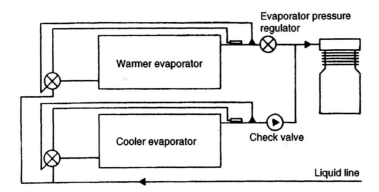

Figure 84 *Location of check valve, multi-evaporator system*

It is designed to stop the flow of water to the condenser when the plant is at rest. An increase in the head pressure will open the valve to allow a greater volume of water through the condenser. A decrease in head pressure will automatically reduce the volume of water flowing through the condenser.

The location of the valve may vary; it may be installed at the inlet to the condenser or at the outlet. Current practice is to install it at the condenser outlet, to ensure that the condenser does not drain when throttling takes place during operation and normal modulation of the valve. Figure 85 shows the valve installed in both positions.

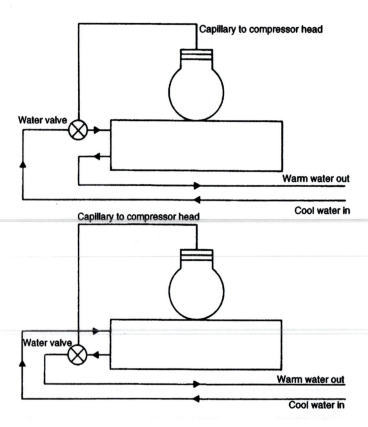

Figure 85   *Location of water regulating valve*

The valves are adjustable. As a guide to operating conditions, the valves should be adjusted to maintain a temperature difference of 7–9 °C (15–18 °F) between the inlet and the outlet water.

Reversing valve

This type of valve is in common use for hot gas defrost systems and room air conditioners. It effectively reverses the flow of refrigerant to the evaporator and condenser when a defrost period is initiated. In room air conditioners, it reverses the flow when heating is required rather than cooling.

The valve is an electromagnetic type and operates by pressure when the solenoid is energized. Figure 86a shows how the valve is installed. When installation of the valve requires brazing, always remove the solenoid coil (which is detachable) and wrap a wet cloth around the valve body to avoid damaging the valve mechanism when intense heat is applied.

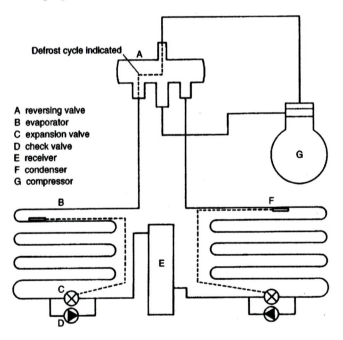

Figure 86a *Installation of reversing valve*

Fusible plugs

Note: Before charging a system with a new or replacement refrigerant check the fusible plug rating.

Condensers and receivers are fitted with fusible plugs. These are safety devices against excessive high side pressures developing in the event of a fire. (Excessive operating pressures are controlled by the high pressure cut-out switch: see Chapter 6.) The plugs are normally brass studs with a 3 mm hole drilled through the centre, which is filled with a low melting point solder.

It is important that, when a replacement is fitted, the correct plug is selected. The following are the required melting points for various systems:

R12   100 °C (210 °F)
R22   76 °C (170 °F)
R502   76 °C (170 °F)
R717   68 °C (155 °F)

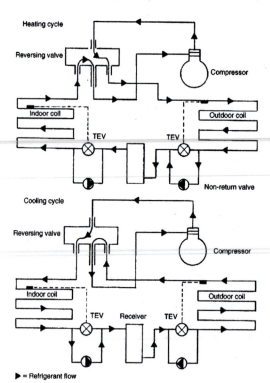

Figure 86b *Heating and cooling cycles for reversing valve system*

12 *Other system components*

This chapter describes the function and operation of a number of components not previously covered in detail.

Solenoid valve

This is an electrically operated magnetic valve. It consists of a coil wound around a sleeve of non-magnetic substance. When energized, the coil carries an electrical current and becomes a magnet.

Inside the sleeve is a movable iron core, which will be attracted by the magnetic field of force to remain suspended in a midway position inside the sleeve. The iron core movement opens the valve. When the coil is de-energized, the iron core returns to its normal position to close the valve.

This type is normally closed. Others are available which are normally open; these close when the coil is energized.

Some solenoid valves have flare fitting unions, whilst others are brazed into position. It is important that the coil is removed when brazed joints are made. For optimum operation the valve should be installed in an upright position and rigidly mounted; the valve should never rely on the system pipework for support.

There are many uses for this type of valve, such as in defrost systems, multiple evaporator systems and pump-down cycles. In effect, it is used wherever it is necessary to stop the flow of refrigerant to a specific section of pipework on a particular application.

Crankcase heater

This is a resistance heater. If externally mounted it is located on the underside of the compressor crankcase. It may be mounted internally as an integral part of the compressor; this is standard for larger commercial equipment.

The presence of liquid refrigerant in a compressor crankcase is wholly undesirable. It could cause dilution of the lubricating oil, resulting in poor lubrication. Liquid refrigerant vaporizing in a crankcase will cause foaming when the compressor operates after an off cycle. Excess oil will then be discharged by the compressor; this could drastically reduce the amount of oil in the crankcase. Non-condensible slugs of oil and liquid refrigerant could enter the compressor cylinders to cause considerable damage to valves, pistons and connecting rods, and even to fracture crankshafts.

Crankcase heaters are essential for compressors installed for low temperature applications, where evaporating temperatures and crankcase temperatures can be extremely low. They are also necessary for remote installations where the compressor is exposed to low ambient temperatures (winter conditions). Whenever the crankcase temperature falls below that of the evaporator, refrigerant vapour will migrate and condense in the crankcase unless a heater is employed to maintain a temperature in the crankcase above the temperature of the refrigerant vapour.

Because of the tendency of oil to absorb miscible refrigerant, a certain amount of refrigerant will always be present in the crankcase.

Check valve

This is sometimes called a non-return valve. It is a simple device used to ensure that fluid or vapour can only travel in one direction and not back up to another part of the system pipework. Check valves have been mentioned in connection with oil separators (Figure 75), multiple evaporators (Figure 84) and hot gas defrosting (Figures 86a and 86b).

Figure 87 shows the construction of the valve, and the locations in a multiple evaporator system.

Sight glass

Two distinct types are commonly used: the clear liquid indicating, and the moisture indicating (Figure 88). Many designs are available, but the common function is to indicate levels and conditions of fluids in the system.

They should be installed close to the liquid receiver. They may be downstream or upstream of the filter drier, but are generally upstream.

Where long liquid line runs or high liquid line risers are necessary, it is advisable to install an extra sight glass immediately before the thermostatic expansion valve. This will indicate the presence of bubbles, possibly due to pressure drop in the liquid line, thereby creating a shortage of refrigerant to the evaporator.

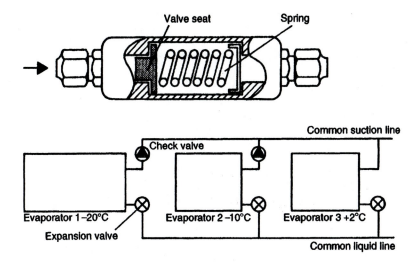

Figure 87 *Check valve construction and location*

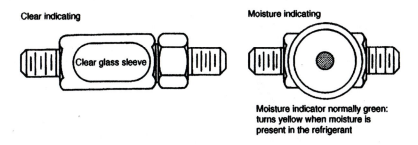

Figure 88 *Sight glasses*

Filter drier

This is installed in the liquid line of the system after the receiver. Construction is generally in the form of a tube which contains coarse and fine mesh filters. These prevent foreign matter such as dirt, metal filings and carbon sludge circulating with the refrigerant. The tube also contains a drying agent or dessicant which will absorb any moisture in the refrigerant (see Figure 89).

A burn-out drier is specifically intended for installation in both the liquid line and the suction line of a system following the replacement of a hermetic or semi-hermetic compressor in which the motor windings have burnt out (see

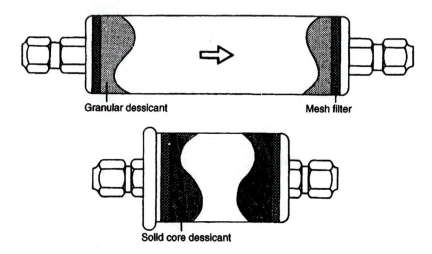

Granular dessicant   Mesh filter

Solid core dessicant

Figure 89   *Filter driers*

Chapter 6). This type of filter drier has the extra ability to retain acids which could be present in the oil residue entrained in parts of the system.

Oil pressure failure switch

This is generally used with compressors incorporating oil pumps and on multiple compressor systems. The function of the control is to stop the compressor(s) when the oil pressure developed by the pump falls below a specific level, or if the oil pressure fails to reach a maximum safe level within a desired period after starting.

The oil pressure, as measured with a gauge, is the sum of the crankcase pressure (suction) and the pressure developed by the pump. The failure switch should be set to operate at the 'useful' pressure, and not at the total pressure. To determine the useful pressure (assuming correct compressor lubrication), subtract the suction pressure from the total pressure. Since the oil pump functions only when the compressor operates, the total pressure will be equal to the crankcase pressure during off cycles.

When the compressor starts, the oil pressure rises to the cut-in point of the switch. The differential switch will open and break the circuit to the heater, and the compressor will operate normally (see Figure 90).

If the useful pressure does not rise to the cut-in point within the time limit (60 to 180 seconds), the differential switch contacts will not open and the

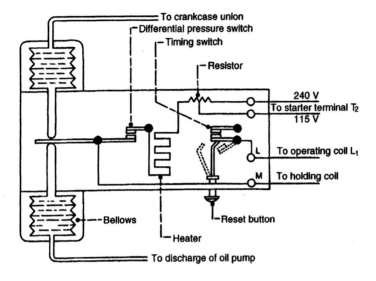

To crankcase union
Differential pressure switch
Timing switch
Resistor
240 V
To starter terminal T₂
115 V
L To operating coil L₁
M To holding coil
Bellows
Reset button
Heater
To discharge of oil pump

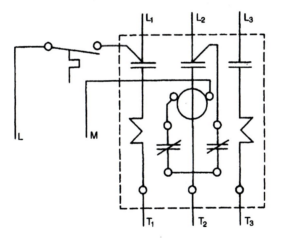

Figure 90 *Typical oil failure switch construction and circuit*

heater will stay in circuit. This causes a bimetal strip in the timing relay to warp and open the timing contacts, which will break the circuit to the starter coil and stop the compressor. Similarly, if the useful pressure falls below the cut-in point during operation, the differential switch will close and energize the heater. The timing relay will stop the compressor after the time delay characteristic of the switch.

Controls are available with 60 and 90 seconds delay, but it must be realized that the time is not variable.

Most current production oil pressure failure switches are provided with terminals for the connection to a crankcase heater. Since the terminal arrangements vary, reference should be made to the wiring diagram provided with the switch.

13 *Drive belts and couplings*

Drive belts

Correct alignment of the flywheel and drive pulley, and correct tensioning of the drive belt(s), are most important during the installation or replacement of belt(s) or motor. Incorrect alignment and tensioning can result in excessive belt wear, loss of compressor efficiency, and motor failure due to faulty motor or compressor drive shaft bearings.

Alignment

This may be carried out with a piece of string or preferably a straight edge (AB in Figure 91), aligning the drive motor pulley to the compressor flywheel. It is essential that the areas X and Y are aligned accurately. Any variation in the pulley or flywheel thickness must be compensated for when aligning the flywheel and pulley faces (area Z).

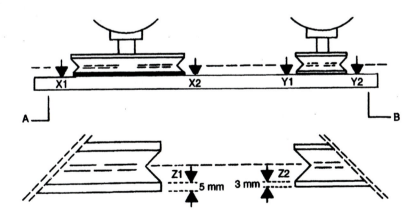

Figure 91 *Belt alignment*

Selection

All drive belts are subject to bending and compressing, which will inevitably result in stretching. Centrifugal force also adds to belt stress. The distance between the flywheel and the pulley plays an important part in the amount of stress and force generated during the operation of the motor and compressor.

It is important, therefore, that drive belts are selected to the design characteristics, i.e. that they are of the correct length, thickness and section. Any increase in centrifugal force will tend to increase the amount of stretch on a belt. A thin belt will create a greater force.

Tension

A loosely fitted belt increases centrifugal force, leads to excessive wear and causes the belt to slip. A belt fitted too tightly may break and have a side effect on shaft bearings.

It is said that a belt is too slack when it jumps off the pulley. However, this is an overstatement because even if the belt remains in place it may still be subject to unnecessary flexing, wear and stress. A recognized method of adjusting the belt tension is to allow a maximum of 25 mm (1 in) deflection at the position shown in Figure 92.

Length

For short belts, that is at up to 1.3 m between shaft centres, the following formula is considered to be accurate enough for belt sizing (Figure 93):

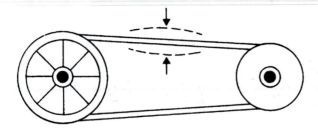

Figure 92 *Belt tension*

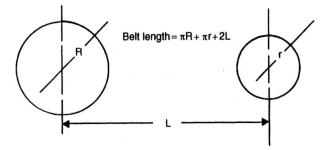

Figure 93 *Belt length*

half circumference pulley + half circumference flywheel + twice distance between shaft centres

$\pi R + \pi r + 2L$

For example, if $R = 225$ mm, $r = 75$ mm and $L = 750$ mm, the belt length is given by

$706.95 + 235.65 + 1500 = 2442.6$ mm $\simeq 2.4$ m

Thickness

Examples of centrifugal force variation with belts of different thickness are shown in Figure 94.

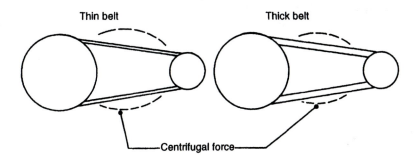

Figure 94 *Belt thickness – centrifugal force variation*

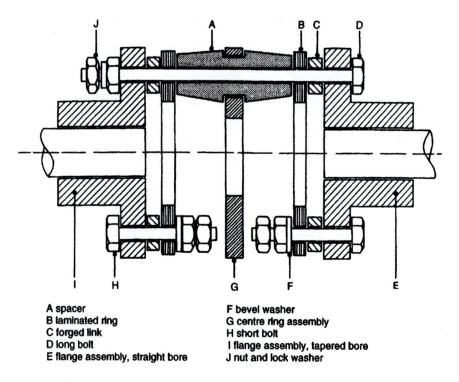

A spacer
B laminated ring
C forged link
D long bolt
E flange assembly, straight bore

F bevel washer
G centre ring assembly
H short bolt
I flange assembly, tapered bore
J nut and lock washer

Figure 95 *Drive coupling*

Drive couplings

Instead of employing a flywheel and drive options, compressors may be designed for direct coupling.

The coupling on small compressors may be of a rigid flange type. Larger compressors will have a more elaborate resilient coupling similar to that shown in Figure 95, consisting of a flange assembly, a centre ring assembly, a number of forged links, and long and short mounting bolts with locking nuts and washers.

Assembly and alignment

Whenever a coupling is fitted, correct alignment is essential to prevent any undue stresses on the compressor, motor shaft and bearings, and to eliminate vibration during operation, which is often at high speeds. Should it be necessary to change a motor or compressor, it is important to note the arrangement of bolts, washers, forged links and nuts during the disassembly, since they must be replaced in the same order.

The most accurate method of aligning a coupling is by using a level indicator gauge, but it can be accomplished with a straight edge and caliper. Both methods will be dealt with in turn. It is advisable to align the coupling to the tolerances stated by the manufacturer of the equipment.

Level indicator gauge

Test for angular misalignment

1 Mount the indicator on the left hand flange as shown in Figure 96, with the stem on the face of the right hand flange.
2 Rotate the equipment, notion the maximum and minimum indicator readings.
3 Move the equipment as necessary to reduce the total indicator reading to 0.0508 mm (0.002 in) or less for each millimetre (inch) of diameter at the indicator stem.

Test for parallel misalignment

1 Set the indicator on the outer surface of the flange as shown in Figure 97.
2 Rotate the equipment, noting maximum and minimum readings.

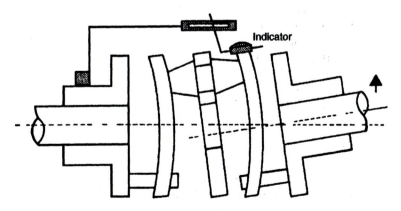

Figure 96 *Angular misalignment*

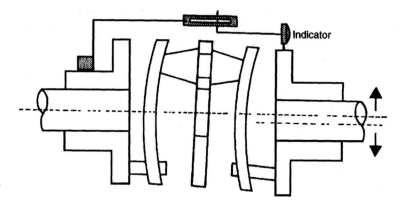

Figure 97 *Parallel misalignment*

3 Move the equipment as necessary to reduce indicator readings to a minimum, taking care not to disturb the setting made in step 1.
4 The coupling should be rotated several complete revolutions to make sure that no endwise creep in the connected shafts is measured.
5 Tighten all bolts. Recheck tightness after several hours operation.

Calipers and straight edge

1 Place a straight edge on the flange rims at the top and sides (Figure 98). When the coupling is in alignment, the straight edge should rest in full contact upon the flange rims.
2 Check the dimension with inside calipers on at least four points of the circumference of the flanges. If these dimensions are within 0.39 mm ($\frac{1}{64}$ inch) of each other, the alignment is satisfactory. Refer to the manufacturer's instructions if available.
3 Move the equipment and repeat steps 1 and 2 if necessary.
4 Tighten all bolts. Recheck tightness after several hours operation.

Appearance

When the coupling is in correct alignment, both laminated ring assemblies will be in a perfect plane at right angles to the shaft centre line.

When the equipment is operating at full speed, both laminated rings should have a distinct and clearly defined appearance when viewed from both the top

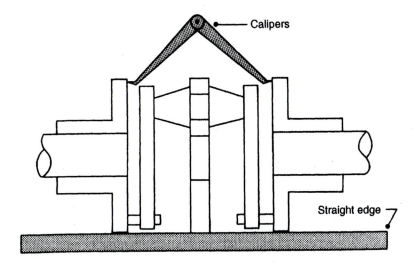

Figure 98 *Alignment using calipers*

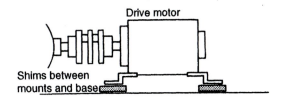

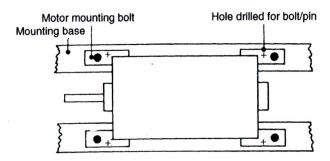

Figure 99 *Alignment using shims*

and the side. Should they have a blurred appearance, the coupling needs to be realigned.

Mounting

During the coupling alignment it may be necessary to raise the level of the drive shaft to coincide with that of the compressor shaft. This can be achieved by placing shims (Figure 99) beneath the motor mounts; the shims are holed for location of motor mounting bolts. It will then be necessary to conduct a parallel alignment test several hours after the equipment has been operating in case the shims have bedded down, thereby putting the coupling out of alignment.

When a coupling has been correctly aligned and operated satisfactorily, holes should be drilled through the motor mounts and mounting base. These can be either tapped to receive a bolt or left untapped for insertion of a dowel pin. Securing will ensure that no angular movement of the driver shaft is possible.

14 *Electrical circuit protection*

Every electrical installation must be protected from current overload. This is achieved by locating a protective device at the commencement of each circuit in the form of a fuse or circuit breaker. Their function is to protect the circuit conductors (cables) and not the appliance or user.

Fuses

Three types of fuse are used (Figure 100); the rewireable or semi-enclosed fuse; the cartridge fuse or fuse link; and the high rupturing capacity (HRC) fuse.

Rewireable fuse

This comprises a fuse holder, an element and a fuse carrier. The holder and carrier can be made of porcelain or bakelite. The fuse holder is colour coded as follows:

45 A	green
30 A	red
20 A	yellow
15 A	blue
5 A	white

This type of fuse is popular for domestic appliances and small commercial units because of cheapness and ease of replacement. It is not recommended for commercial refrigeration duty because of these disadvantages:

1 The fuse carrier can be loaded with the wrong size fuse wire.
2 The fuse element tends to weaken after long usage owing to oxidation of the wire by heating in air. This causes it to fail under normal conditions, i.e. normal starting current surges are sensed by the fuse as an overload.

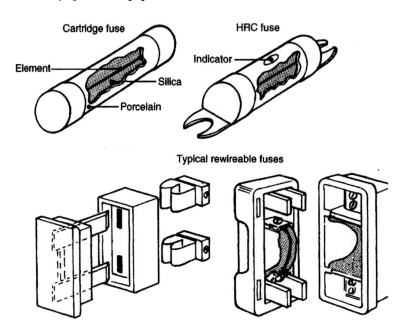

Figure 100   *Fuses*

3 The fuse holder and carrier can be damaged as a result of arcing in the event of an overload.

Cartridge fuse

This consists of a porcelain tube with metal end caps to which the fuse element is attached. The tube is filled with silica.

These fuses may be used in plug tops with 13A socket outlets and in distribution boards. They are recommended for refrigeration duty. They have the advantages over the rewireable types of not deteriorating, of being more accurate in breaking at the rated value, and of not being subject to arcing.

HRC fuses

This is a sophisticated version of the cartridge fuse. It is normally used for the protection of motor circuits in commercial and industrial installations. It consists of a porcelain body filled with silica and with a silver element; the body terminates in lug-type end caps.

These fuses are fast acting, and can discriminate between a starting surge and an overload. An indicating element shows when the fuse is ruptured.

Rating

The selection of fuse ratings depends on the full load current, the locked rotor current and the cable size. The current ratings for tinned copper wire are shown in Table 6.

Fusing factor

Different types of fuse provide different levels of protection. Rewireable fuses are slower to operate and are less accurate than cartridge types.

To classify the protection devices it is important to know the fusing performance. This is achieved by the use of a fusing factor:

$$\text{fusing factor} = \frac{\text{fusing current}}{\text{current rating}}$$

Here the fusing current is the minimum current causing the fuse to rupture, and the current rating is the maximum current which the fuse can sustain without rupturing. For example:

1   A 5 A fuse ruptures only when 9 A flows; it has a fusing factor of 9/5 or 1.8.

Table 6   *Current ratings for tinned copper wire*

Rating A	Size mm	Rating A	Size mm
3	0.15	30	0.85
5	0.20	45	1.25
10	0.35	60	1.53
15	0.50	80	1.80
20	0.60	100	2.00
25	0.75		

2   The current rating of a cartridge fuse is 30 A and the fusing factor is 1.75; the fuse will rupture at $30 \times 1.75 = 52.5$ A.
3   The current rating of an HRC fuse is 20 A and the fusing factor is 1.25; the fuse will rupture at $20 \times 1.25 = 25$ A.

It must therefore be realized that a fuse is rated at the amount of current it can carry, and not the amount at which it will rupture. Rewireable fuses have fusing factors of approximately 1.8; cartridge fuses of between 1.25 and 1.75; HRC fuses of up to 1.25 maximum; and motor cartridge fuses of 1.75.

Circuit breakers

A fuse element provides protection by destroying itself and must be replaced. It cannot be tested without destruction; therefore the result of a test will not apply to the replacement.

The circuit breaker (CB) is an automatic switch that will open in the event of excess current and can be closed again when a fault is rectified. The switch contacts are closed against spring pressure, and held closed by a form of latch arrangement. A slight movement of the latch will release the contacts quickly under the spring pressure to open the circuit; only excessive currents will operate it.

Two types exist: thermal and magnetic.

Thermal

The load current is passed through a small heater, the temperature of which depends upon the current it carries. The heater will warm up a bimetal strip. When excessive current flows the bimetal strip will warp to trip the latch mechanism.

Some delay occurs owing to the transfer of heat produced by the load current to the bimetal strip. Thermal trips are suitable only for small overloads of long duration. Excessive heat caused by heavy overload can buckle and distort the bimetal strip.

Magnetic

The principle used in this type is the magnetic force of attraction set up by the magnetic field of a coil carrying the load current. At normal currents the

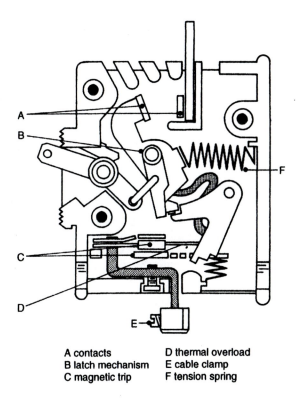

A contacts   D thermal overload
B latch mechanism   E cable clamp
C magnetic trip   F tension spring

Figure 101   *Typical magnetic circuit breaker*

magnetic field is not strong enough to attract the latch. Overload currents will increase the force of attraction and operate the latch to trip the main contacts.

A typical magnetic circuit breaker is shown in Figure 101.

15 *Drive motors*

Before installing or replacing a drive motor, it must be established that the voltage and frequency of the motor (as stated on the nameplate) agree with those of the supply.

A motor must not be connected to a supply outside the range stated on the nameplate. If connected to a low voltage supply it may not start. If it does start it may not run to its full speed; the windings will then overheat, and may burn out if continued starting attempts are made by the action of the overcurrent protector.

It is therefore important that the overload protector is of the correct rating if a replacement is made. Those fitted to open drive motors should be rated as near as possible to 125 per cent of the nameplate current.

Installation details

Earthing

The motor frame must always be connected to the local earth circuit. It is most important to ensure that the flexible earthing strip between the motor frame and the cradle is not broken and that the correct connection is made.

When water cooled systems are involved, under no circumstances must the water supply tubing be used for earthing.

Lubrication

Motor bearings should be lubricated at the time of installation or replacement. For sleeve bearings use high grade light mineral oil; for ball bearings use pure mineral grease.

Do not apply an excess of oil or grease, because excess lubricant can damage motor windings and switch mechanisms. Do not allow oil to contact the resilient rubber motor mounting pads. A small drop of oil should be applied to the swivel pivot.

Bearings

Worn bearings will probably be evident in the form of a hum or rattle from the motor.

The end play should not exceed 0.25 mm. If shims are used to correct the clearance or take up wear, they must not be fitted at one bearing only. This will result in the rotor being out of centre with the stator and will cause humming.

Thermal overloads

These are usually self-resetting after cooling. They are set to cut out when the temperature of the motor windings reaches approximately 90 °C (194 °F).

Nuisance tripping of a motor may be the result of poor ventilation of the motor or the load on the compressor. This should be checked before any replacement is made.

Direction of rotation

To reverse the rotation of a capacitor start motor, the direction of current flow in either the start or the run winding must be reversed (but not both windings). This entails reversing the internal connections of a winding, as shown in Figure 102. For three phase motors it is only necessary to reverse any two electrical leads to the motor terminals.

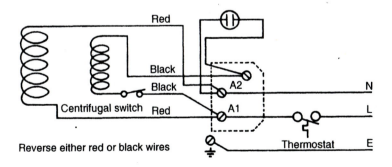

Figure 102   *Reversing direction of rotation*

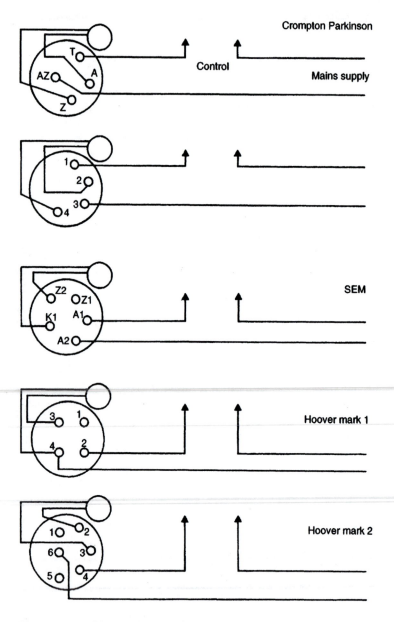

Figure 103 *Drive motor connections*

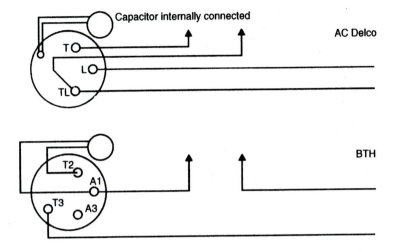

Figure 103 *(continued)*

Terminal board connections

Suppliers of drive motors are numerous and the terminal identification is not standard. A variety of connections for motors in common use is provided in Figure 103.

Three phase electrical connections

Smaller compressor motors incorporate overload protectors, either internally or externally mounted, and use various small starting devices. By contrast the three phase motor, being larger, requires more sophisticated methods of automatic starting and overload protection, provided by contactors or starters.

Contactors and starters

A contactor is a power operated switch suitable for frequent making and breaking of an electrical circuit. A starter is a power operated switch with inbuilt overcurrent features, employed for starting a motor and protecting it during running. Both controls enable large currents to be switched by means of a system control (thermostat) with low current rating.

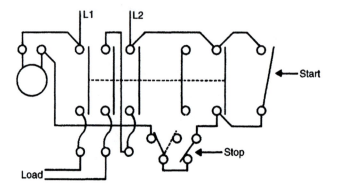

Figure 104   *Electrical connections for a single phase starter*

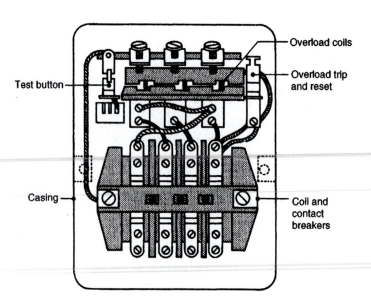

Figure 105   *Type of three phase starter*

Suppliers of these controls are numerous. Each has its own options with regard to overload protection, single phasing protection and accessibility for component replacement. Wiring diagrams for the controls are normally supplied or are available on demand. These should be referred to before any attempt is made to install, make electrical connections to or diagnose faults on the controls. Because of the different types of control, only typical examples are dealt with here. Figure 104 shows connections for single phase starting, and Figure 105 shows a particular three phase starter.

Direct-on-line starter

With the standard squirrel cage induction motor, it is advantageous to have DOL starting. This allows a cheaper motor starter and control gear to be used, and also imposes less strain on motors during starting.

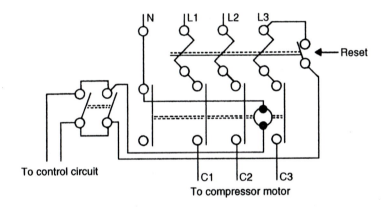

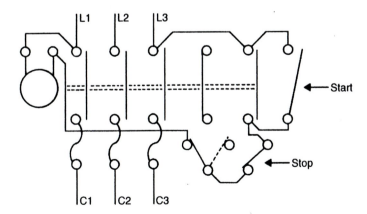

Figure 106 *Electrical connections for three phase DOL starters*

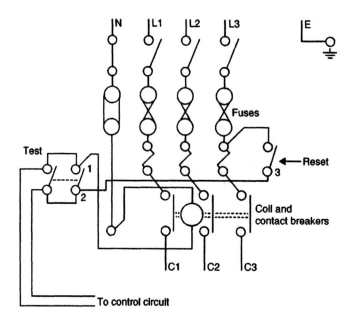

Figure 106 *(continued)*

However, electrical supply authorities will normally restrict starting currents. DOL starters should only be specified for motors within the supply authority limits.

Figure 106 shows the electrical connections for three phase DOL starters.

16 Commissioning of refrigerating systems

Commissioning can be defined as the advancement of an installation from the stage of static completion to full working order to specific requirements.

A number of different contractors may have been involved: refrigeration fitters, electricians, plumbers, builders, carpenters and shop fitters. Close cooperation among all concerned is essential. It is therefore desirable that the commissioning of a plant be carried out under the guidance of a single authority; that usually means the service or installation engineer.

The customary safety procedures must be observed at all times. During commissioning the following will prevail:

1 Use only the specified refrigerant (manufacturer's data).
2 Substitute refrigerants should only be used if approved by the equipment manufacturer.
3 Do not re-use refrigerant which might be contaminated.

Contaminants

During installation the system should be kept as clean and dry as possible, with the least exposure to air. Avoid the entry of foreign matter such as solder fluxes, solvents, metal and dirt particles, and carbon deposits; the last are the outcome of soldering joints without passing a neutral atmosphere through the pipework.

Failure to take precautions may result in corrosion caused by air and moisture, or by an oxidizer under high temperature conditions. Other problems include:

1 Copper plating due to contaminated oil. It forms on bearings and planed surfaces in high temperature areas. Moisture in the system can also be the cause.

2   Freezing, which may take place in the system components if the correct dehydration procedure (evacuation) is not adopted.
3   Sludging, that is the chemical breakdown of oil under high temperature conditions in the presence of non-condensables.

The chemical breakdown or thermal decomposition of both refrigerant and oil in temperatures in excess of 150 °C (300 °F) is more likely to occur with refrigerants R22 and those in the R500 group. In the presence of hydrogen-containing molecules the thermal decomposition produces hydrochloric and hydrofluoric acids, a condition which is wholly undesirable if hermetic or semi-hermetic compressors are employed. For this reason it is imperative that the evacuation period is adequate.

Evacuation

It is imperative, with halogen refrigeration systems, that all traces of air, non-condensibles and moisture be removed. If this is not achieved then the presence of air or non-condensibles will cause abnormally high discharge pressures and increased temperatures, resulting in the conditions relating to high operating pressures previously explained.

Air in the system will also mean that a certain amount of moisture contained in the air will be circulated with the refrigerant under operating conditions. This moisture could freeze at the orifice of an expansion valve or liquid capillary to prevent refrigerant flow to the evaporator, should the filter drier become saturated.

When a system has been pressure leak tested, traces of nitrogen may also be present to further aggravate the high discharge pressure condition.

There are two ways in which a system can be evacuated, the deep vacuum method and the dilution method.

Deep vacuum method

To meet the requirements of a contaminant-free system, a good vacuum pump is necessary. Under normal ambient temperatures a vacuum of 2 torr should be achieved with a single deep-vacuum cycle.

The length of a deep-vacuum cycle can vary considerably: the larger the installation, the longer the cycle. This may be left to the discretion of the commissioning engineer, as stipulated by company policy, or a specific period may be requested by the customer. Obviously a large high vacuum pump will expedite the procedure. It is not unusual for a system to be left on vacuum for 24 or 48 hours, or even for several days, to ensure that it is completely free from contaminants.

The advantages of a deep vacuum are that (a) there will not be any appreciable loss of refrigerant other than the final trace charge administered while leak testing, and (b) it is possible to reclaim a trace charge of refrigerant from a large system (see Chapter 16 relating to contaminants and refrigerant recovery). Also, the immediate environment will not be polluted by refrigerant vapour so that it is difficult to carry out a final leak test when the system is charged. This will be evident when comparison is made with the dilution method.

Dilution method

The dilution method or triple evacuation should be carried out using OFN (oxygen free nitrogen) and **not with a trace charge**.

1   The initial nitrogen charge should be left in the system for at least 15 to 30 minutes. It can then be re-evacuated to a vacuum of 5 Torr.
2   This vacuum is then broken with another OFN charge allowing time for it to circulate the system.
3   Re-evacuate and charge the system with refrigerant.

This repetition may appear to be unnecessary but after a single or double evacuation small pockets of non-condensables may still be entrained in the system pipework or controls. By repeatedly breaking the vacuum with OFN these pockets will be dispersed or diluted by the OFN.

After each evacuation the pump should be switched off and, after a few minutes settling period, a vacuum reading taken. The system should then be left for another 30 minutes and another reading taken. A rise in pressure means that there is still a certain amount of moisture present.

Under no circumstances should the system compressor be used for the evacuation of the system.

A comparison of vacuum gauge graduations is given in Figure 107. Note that 1 torr = 1 mm Hg = 1000 μm Hg and micrometres are referred to as microns.

Figure 108 shows a typical arrangement for connecting a vacuum pump for deep evacuation.

During the evacuation of the system the evaporator fan(s) may be operated and defrost systems switched to the heating cycle in order to raise the temperature in the evaporator. Heaters must not remain energized for excessive periods in case of overheating of the evaporator and possible damage. It

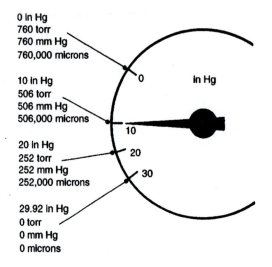

Figure 107   *Vacuum pump gauge*

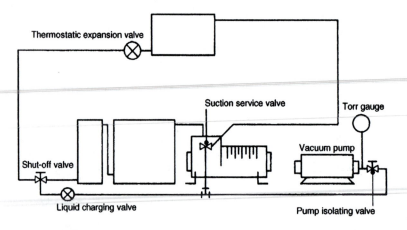

Figure 108   *Typical deep vacuum pump arrangement*

is also very important to ensure that no parts of the system are isolated from the vacuum pump.

Figure 109 shows a triple evacuation arrangement. When the pump is operating, the isolator valve must be open, the service valves on the compressor in the midway position, the liquid shut-off valve at the receiver open, and the refrigerant cylinder valve closed. Both valves on the gauge manifold must

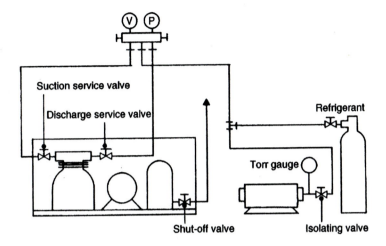

Figure 109 *Triple evacuation arrangement*

be open. When breaking the vacuum with refrigerant vapour pressure, ensure that the pump isolating valve is closed.

Table 7 shows the pressure/temperature relationship for water. When evacuating a system, remember that there must be an adequate temperature difference between the ambient temperature and that of the water to provide the heat necessary to vaporize the water.

Commissioning checks

When the system has been evacuated, the next step is to carry out a thorough leak test. Methods for this have been dealt with in Chapter 3.

The system may then be charged with refrigerant. Irrespective of which method of charging is used, it must be remembered that the compressor is being operated for the first time. It will have been standing for a considerable period, and oil will have drained from cylinders and bearings.

Before attempting to start the compressor the following checks must be made:

1 Ensure that the electrical supply voltage is correct according to the compressor motor nameplate.
2 Check the voltage at the compressor motor terminals.
3 Check the fuse ratings.
4 Set the safety controls and check operation.

Table 7   *Pressure/temperature relationship for water*

Water boiling point °C	N/m²	Absolute pressure bar	torr (mm Hg)
100	101356	1.01	760
93.3	79499	0.795	596.11
65.6	29574	0.296	221.8
37.8	6550	0.066	49.11
26.1	3379	0.034	25.34
21.1	2503	0.025	18.8
15.6	1724	0.017	12.93
12.8	1476	0.015	11.07
10.0	1227	0.012	9.2
7.2	1014	0.010	7.6
4.4	841	0.008	6.3
1.7	690	0.007	5.2
0	614	0.006	4.6
−1.1	558	0.005	4.2
−3.9	441	0.004	3.3
−6.7	345	0.0035	2.6
−9.4	276	0.0028	2.07
−12.2	213	0.0020	1.6
−15.0	172	0.0017	1.3
−17.8	124	0.0010	0.93

101,356 N/m² = 760 torr; 133.4 N/m² = 1 torr

5   If the compressor is an open type, turn it over by hand if this is practical to verify free rotation.

At this stage it is also advisable to prepare an equipment log and to record electrical data, control settings, operating pressures and temperatures noted during commissioning, in case of a break in continuity during this procedure and for future reference.

Oil addition and removal

When the system has been fully charged and the controls set, the equipment is operated at average evaporating conditions. The compressor oil level must then be checked.

All new compressors receive an operating oil charge during manufacture, but this does not allow for oil trapped within components and controls and circulating with the refrigerant. This is the reason why oil levels must be checked and oil added to the system at the time of installation or commissioning.

Adding oil

Only the make and grade of oil specified by the manufacturer of the compressor must be used. Oils have special characteristics for pressure, temperature and load conditions, especially those used for low temperature applications.

The larger compressors will have oil-level-indicating sight glasses located in the crankcase. If compressors are used in parallel, there is sometimes a sight glass in the middle of the oil equalizer line. During operation the oil level in the crankcase may well fluctuate. It is generally accepted that a level maintained at one-third to one-half of the way up the sight glass is satisfactory.

Before any oil is added, the system must be fully charged. This is most important when the compressor is located above the level of the evaporator. Oil levels should be rechecked after a system has completed its initial pulldown, or has operated for at least two hours. If a sight glass indicates foaming, ensure that this is not the result of absorbed refrigerant. Systems prone to this condition will probably be fitted with crankcase heaters.

Oil may be added to the crankcase of larger compressors simply by pumping down and reducing the crankcase pressure to enable oil to be poured when the filler cap has been removed. However, filler holes tend to be small on some compressors and an oil pump may be used.

Too much oil in a crankcase can cause damage to a compressor by creating a dynamic pressure during operation.

The procedures for adding oil will obviously vary according to the type of compressor.

Charging pump and simple filling

The oil charging pump is similar to a cycle pump and needs no explanation. The correct oil level must be attained, and reference should be made to the manufacturer's data.

Sometimes it suffices to add oil to a compressor until it runs out of the filler hole; a dipstick will be required for other types. In each instance the oil level will be such that it is approximately 25 mm (1 inch) below the crankshaft so that bearings or splashers immerse in the oil as the shaft rotates (Figure 110).

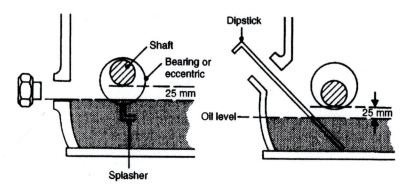

Figure 110   *Manual oil filling*

Vacuum pump

When a compressor has a sight glass it is a simple task to add oil. The procedure is as follows (Figure 111):

1   Pump down the system to reduce the pressure in the crankcase to 0.1 bar (1 psig) and front seat both suction and discharge service valves.
2   Remove the oil filler plug and fit a charging line incorporating a shut-off valve and an adaptor to insert into the filler hole.
3   Place the free end of the charging line into a container of clean and uncontaminated oil from a sealed can. Crack off the suction service valve from the front seat position and raise the crankcase pressure to 0.1 bar.

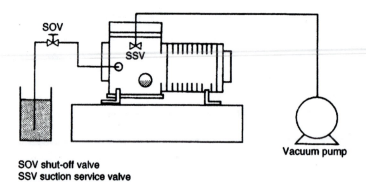

SOV shut-off valve
SSV suction service valve

Figure 111   *Vacuum pump method of adding oil*

Open the shut-off valve slowly and purge the charging line through the oil in the container. Front seat the suction service valve.

4   Connect a vacuum pump to the gauge union of the suction service valve.
5   Switch on the vacuum pump and reduce the pressure in the crankcase to slightly below atmospheric, allowing the oil to be drawn in until the correct level is reached.
6   Stop the vacuum pump, crack off the suction service valve from the front seat position, purge oil from the charging line and close the shut-off valve. Then front seat the suction service valve.
7   Remove the charging line and replace the oil filler plug.
8   Fully back seat and crack off both service valves, or set to operating positions.
9   Leak test the compressor.
10 Start the system and allow it to settle down to average operating conditions. Recheck the oil level.

When charging oil, ensure that the charging line is always below the oil level in the container.

Compressor charging

When the compressor design is such that a suction strainer and oil return to the crankcase is featured, oil may be added via the suction service valve gauge union in much the same way as described above but using the compressor to draw a vacuum instead of a vacuum pump (Figure 112).

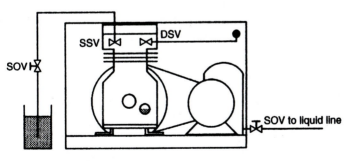

SSV suction service valve
DSV discharge service valve
SOV shut-off valve

Figure 112   *Using compressor to draw vacuum*

Draining compressor oil

This may be necessary when too much oil has been added, when maintenance contracts stipulate a periodic oil change, or when a system has been contaminated.

Assuming that the compressor does not have an oil drain facility and the removal of the base plate or sump plate is impractical or uneconomical, two simple methods may be adopted.

Vacuum method

This requires an air tight container or preferably a graduated cylinder so that the amount of oil removed can be measured and the precise amount replaced.

By pulling a vacuum on the container or cylinder, the oil will be drawn out of the compressor into the cylinder (Figure 113). During this process the compressor must be isolated from the system by front seating the service valves.

System pressure method

With a length of tubing and an adaptor fitted to the filler hole after pressure in the crankcase has been reduced and both service valves front seated, create a positive pressure in the crankcase by cracking off the suction service valve from the front seat position (Figure 114). Provided the tubing reaches to the bottom of the crankcase or sump, the oil will be forced out of the compressor and into the container.

Correcting the oil level

The pressure method above can be used to reduce the level of oil in the crankcase in the event of an overcharge. By fitting an adaptor tube which terminates at the correct level below the crankshaft, oil will be forced out of

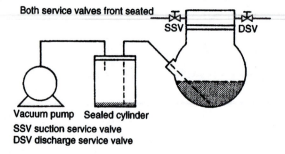

Both service valves front seated

SSV   DSV

Vacuum pump   Sealed cylinder
SSV suction service valve
DSV discharge service valve

Figure 113   *Vacuum oil draining*

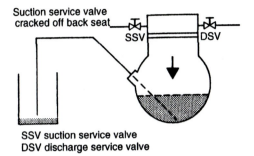

Suction service valve
cracked off back seat

SSV   DSV

SSV suction service valve
DSV discharge service valve

Figure 114 *System pressure oil draining*

the compressor when the crankcase is pressurized, but only until the oil level reaches the end of the tube.

Environmental impact of CFCs

Chlorofluorocarbons (CFCs) are chemical compounds which have been developed for use as refrigerants. Their molecular structure is based on either methane or ethane; one or more of the hydrogen atoms is substituted by chlorine or fluorine.

The CFC refrigerants soon replaced most other refrigerants except ammonia, which is still in use today. Other products made from CFCs were then used for aerosols, expanded foam processes and degreasing agents.

One desirable property of the compounds which appealed to manufacturers of refrigerants is their chemical stability. It is this long term stability which contributes to the pollution of the atmosphere. Once released, CFCs remain in the atmosphere for years. At low altitudes this does not present a problem. However, when they reach the upper atmosphere CFCs, like ozone, are broken down by ultraviolet light. This results in the release of free radical chlorine atoms, which interfere with the normal formation of ozone and contribute to the greenhouse effect (about 10–15 per cent).

There are now restrictions imposed upon manufacturers of certain CFCs by the Montreal Protocol, an international agreement which came into force in January 1989. Within ten years, production of CFCs should have been reduced to 50 per cent of the 1986 levels.

Refrigeration service and installation engineers can assist in reducing the emission of CFCs. Design engineers can ensure that systems are constructed so that emissions are minimal when various forms of maintenance or repairs are necessary.

Good refrigeration practice for CFC systems

It is a well known fact that an engineer required to work on an ammonia system will be very careful not to allow undue discharges of refrigerant because of its toxicity. Since CFCs are non-toxic it has been common practice to discharge them to atmosphere. This must now be regarded as malpractice and cease forthwith.

Evacuation

It was also past practice to use the dilution or triple evacuation method, in which a small amount of refrigerant is used to dilute the atmosphere within the system, discharging between evacuations. This method can still be employed, but instead of releasing the refrigerant to atmosphere it should be reclaimed by decanting to a cooled refrigerant cylinder.

Alternatively, the more practical deep evacuation method should be adopted.

Cleaning condensers

The practice of cleaning condensers with a refrigerant should be discontinued. Proprietary brands of degreasing and cleansing agents which are environment friendly are readily available.

Should it be necessary to apply pressure to ensure penetration to fins and pipework, then nitrogen should be used.

Decanting refrigerant

Special care must be taken not to overfill the cylinder, and to use only those which are free from any contamination by oil, acid or moisture.

Do not mix grades of refrigerants. Always use a cylinder for the specific refrigerant for which it is designated.

Removal of refrigerant from sealed systems

This can be achieved by fitting a line tap valve to the system and connecting to a recovery cylinder.

Removal of contaminated refrigerant

Contaminated refrigerant which may have resulted from a compressor burnout or a water cooled condenser leak must not be used to recharge the system. It must be recovered and sent away for reprocessing or disposal.

It should be decanted into special recovery cylinders available from the manufacturers or suppliers of refrigerants. *Never* use service cylinders for reclaiming contaminated refrigerants.

When removing refrigerant charges or decanting from systems, adequate protective clothing and goggles must be worn. All safety procedures must be observed.

It is advisable to include an isolating hand shut-off valve at the cylinder end of the charging line to minimize purging. Ensure that the valve is open before discharging from the system. Use charging lines without the Schraeder inserts.

Refrigerant recovery system

Figure 115 typifies a portable refrigerant recovery unit now in use within the refrigeration industry.

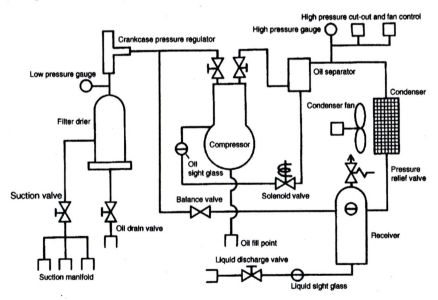

Figure 115 *Schematic layout of a refrigerant recovery unit*

The compressor is an open reciprocating type, driven by a 1.5 kW single phase drive motor. An oil separator is fitted in the discharge line. Oil return is by float control and a solenoid operated return valve, which is energized by a delay timer.

A conventional finned air cooled condenser has a 75 watt fan motor. A high pressure control is set to cycle the fan at 150 psig (10 bar) cut-in and 120 psig (8 bar) cut-out. The fan control is independent of the compressor control and will continue to run when the compressor stops.

The receiver, capable of holding a relatively large volume of liquid refrigerant, is fitted with a sight glass. This indicates that the receiver is full when the liquid level is half-way up the glass. A pressure relief valve is also fitted to the receiver and set at 400 psig (26.5 bar). A liquid discharge valve permits drainage of the receiver through a standard hose connection.

A crankcase pressure regulator is installed to manually control the compressor suction to give 9.5 amperes maximum load on the drive motor. A suction valve isolates the suction manifold from the recovery unit. The manifold has three hose unions for multiple connection to various parts of the system from which the refrigerant is to be recovered.

A balance valve connected to both the high and the low sides of the system can be opened to allow:

1 The unit to be operated on a simulated load in order to set the motor load current.
2 The compressor to be started unloaded.
3 Ease of unit evacuation when changing refrigerants.
4 Manual air purging.

Standard glycerine dampened compound and pressure gauges measure the system pressures.

A suction filter drier contains a burn-out core, and acts as a suction boil-off reservoir should small amounts of liquid refrigerant be passed through the unit. It also allows any oil removed from the refrigerant to be drained via a suction filter drain valve with a standard hose connection.

Part Three

Replacement Refrigerants and Ozone Depletion

17 *Ozone depletion*

The refrigeration industry has been aware of ozone depletion and global warming in so much as it has, during the late 1980s, paid particular attention to the ozone depletion. As a result atmospheric science and control regulations of undesirable substances have led to the introduction of a new range of refrigerants. The environmental issue of global warming became more important. Global warming results from atmospheric phenomena commonly described as the 'greenhouse effect'. To simply describe global warming it has been determined that infrared energy radiated from the earth's surface is absorbed by certain atmospheric gases. These gases are classified as greenhouse gases or GHGs. The GHGs allow solar energy to penetrate the atmosphere but do not permit the resultant infrared radiation from the earth to be released. The long term effect of this is a gradual increase in the atmospheric and surface temperature of the earth.

Numerous natural and man-made gases which can be classified as GHGs are evident. The major contributors to global warming are as follows:

Carbon dioxide	50–60%
Methane	10–20%
CFCs, HCFCs, HFCs	20%
Others	10%

The Montreal Protocol

Once the effects of CFC gases on the ozone layer were fully appreciated, leading industrialized countries negotiated a common policy to limit the use of these gases. The term CFC is short for chlorofluorocarbon and refers to a number of compounds developed in the late 1920s. These compounds have a range of qualities to enable them to be used as refrigerants. The term HCFC is short for hydrochlorofluorocarbon and the term HFC is short for hydrofluorocarbon.

The Montreal Protocol as it is now known was signed by 49 countries in September 1987. This international agreement which came into force in January 1989 imposed restrictions upon manufacturers of CFCs.

The main features of the Protocol are:

1   Six months after coming into force in mid-1989 consumption should not exceed 1986 levels.
2   By 1 July 1993 consumption should not exceed 80% of 1986 levels.
3   By 1 July 1998 consumption should not exceed 50% of 1986 levels.

The controlled substances are:

CFCs 11, 12, 113, 114, and 115. This will also affect CFC 500 which is a blend of CFC 12 and CFC 115.

When a CFC gas is prefixed by the letter R it denotes that the CFC is being used as a refrigerant.

The greenhouse effect

The infrared energy radiated from the earth's surface by other atmospheric gases is reflected by dense molecules of the greenhouse gases in the upper atmosphere being retained in the atmosphere. The retention of this energy in sufficient quantity will result in global warming. An increase in world temperature of only a few degrees will produce dramatic climatic changes and a rise in sea levels.

CFCs are known to account for some of this problem, but less when compared with other greenhouse gases such as carbon dioxide which is recognized as the greatest contributor.

Carbon dioxide accounts for 50–60% of greenhouse gases. Methane, the next ranking contributor at 10–20%, is largely the product of normal natural processes, for example marsh gas, and is more difficult to control.

Ozone depletion

Scientists believe that an increase of ultra violet light reaching the earth is associated with skin cancers. Plants are sensitive to sunshine and too much UV light could destroy them and most at risk are cereals. Scientific research has also established that UV radiation levels have a serious effect on Antarctic phytoplankton. Any drastic reduction of these single cell plants will be breaking the first link in the food chain for all southern oceans.

In addition to the CFCs used as refrigerants aerosols came under scrutiny and now contain environmentally 'friendly' substances.

Table 8

CFC number	Chemical formula	ODP
R11	CCl_3F	1.0
R12	CCl_2F_2	1.0
R113	$C_2Cl_3F_3$	0.8
R114	$C_2Cl_2F_4$	0.8
R115	$C_2Cl_1F_5$	0.4
R500	Azeotrope	0.73
R502**	Azeotrope	0.23
HCFC22*	$CHCl_2F_2$	0.05

* Not controlled by the Protocol.
** Contains a controlled substance.

It could be said that refrigerators and commercial refrigerating equipment are a major contributor. It is more difficult to legislate against the use of CFCs in these appliances than to ban their use in aerosols. It is not as simple as just replacing the CFC with an alternative 'safe' substance. The amount of CFCs released to the atmosphere from refrigerators is small in comparison to that from commercial refrigerating equipment. The real problem is related to leakages in the systems and during servicing when old units are scrapped for whatever reason. Today there is a code of practice to follow and harmful gases must be reclaimed. This puts a greater responsibility upon the service and installation engineer.

Refrigeration vapour compression systems have in recent years been designed and installed for refrigerants R12, R22 and R502. These have now been replaced and the reason for this is the ozone depletion potential (ODP) of these refrigerants. Various refrigerant suppliers have published information in respect of ODP rates and examples are given in Table 8.

HCFC22 offered a solution for replacing the CFCs in a great number of refrigeration systems in the short term, its use being permitted for service and maintenace after it was banned from use in new equipment.

18 New and replacement refrigerants

For the long term replacement of CFCs and HCFCs, various manufacturers have developed products designed to meet the requirements of the industry.

In October 1990 ICI was the first company to introduce HFC 134a, an alternative to CFCs used in refrigeration and air conditioning. Since then other refrigerants have been introduced and marketed by ICI as Klea 134a.

Klea 32, the second in the Klea range, went into production in 1992. As these refrigerants were successful a number of blends were introduced to the range including Klea Blend 60 and Klea Blend 66, ozone benign refrigerants based on Klea 32 and Klea 134a. These are the alternatives to HCFCs. Each of these is non-toxic, non-flammable and has a good heat transfer property. All have undergone extensive trials, both in laboratory and systems in the field.

These refrigerants have a zero ODP and are currently not subject to regulation. European Union Regulation member states agreed to cease producing CFCs with effect from 1 January 1995. Regulations regarding the HCFC products meant that they would eventually be phased out although it was understood that interim products were necessary to assist the refrigeration industry during transition away from CFC refrigerants.

Tables 9 to 12 list the replacement and new refrigerants as long term CFC and HCFC replacements together with interim products.

There exists some confusion with regard to CFCs, HFCs and HCFCs; also refrigerant blends appear to come under various categories. In some quarters it is believed that CFCs are 'banned'. This is not exactly correct; they are still obtainable for those willing to pay the price and are still widely used. The manufacture of CFCs has now ceased.

There are a great number of refrigeration systems charged with CFCs operating nationwide and users are unlikely to go to the expense of changing the refrigerant charge until a major breakdown occurs.

To summarize: by the year 2000 or possibly earlier R22 will be phased out, meanwhile the refrigerants listed in Table 13 may be used.

Table 9

ASHRAE no.	ICI name	Composition (% w/w)			Replaces
		R32	R125	R134a	
R134a	Klea 134a	—	—	100	R12
R407a	Klea 407A (Klea 60)	20	40	40	R502
R407B	Klea 407B (Klea 61)	10	70	20	R502
R407C*	Klea 407C (Klea 66)	23	25	52	R22
R32	Klea 32	100	—	—	
R125	Klea 125	100	—	—	

* Not controlled by the Protocol.

Table 10 *Refrigerants based on components other than R32, R125 and R134a*

R508*	Klea 5R3	23/116	R503
R404A	R404A	125/290/134a	R502
R402A	Arcton 402A	22/290/125	R502

* Contains a controlled substance.
ASHRAE have provisionally allocated these numbers.

Table 11 *Chemical names and formulae for SUVA refrigerant*

Registered trademark	Chemical name	Formula	Boiling point (°C) at 1 bar
SUVA Centri-LP HCFC 123	2, 2,-Dichloro, 1, 1, 1, Trifluoro ethane	$CHCL_2CF_3$	27.9
SUVA Chill-LP HCFC 124	2, Chlor-1, 1, 1 2-Tetrafluoro ethane	$CHCLFCF_3$	−11.0
SUVA Freez-HP HFC 125	Pentafluoro ethane	CHF_2CF_3	−48.5
SUVA Cold-MP HFC 134a	1, 1, 1, 2 Tetrafluoro ethane	CH_2FCF_3	−26.5
HFC 23	Trifluoro	CHF_3	−82.0

Table 12 *Chemical names and formulae for FREON refrigerants*

Registered trademark	Chemical name	Formula	Boiling point (°C) at 1 bar
FREON 11	Trichlorofluoromethane	CCl_3F	23.8
FREON 12	Dichlorofluoromethane	CCl_2F_2	-29.8
FREON 22	Chlorodifluoromethane	$CHClF_2$	-40.8
FREON 500	Azeotrope of dichloro-fluoromethane and 1,1-difluoroethane	CCl_2F_2/CH_3CHF_2 73.8%/26.2%	-33.5
FREON 502	Azeotrope of chloro-difluoromethane and chloropentafluoroethane	$CHClF_2/CClF_2CF_3$ 48.8%/51.2%	-45.4

Table 13

HCFC retrofit blends with R22 (transitional alternatives)

R401a	R22/R152a/R124	Replaces R12
R401b	R22/R152a/R124	Replaces R12
R409a	R22/R142b/R124	Replaces R12
R402a	R22/R125/R290	Replaces R502
R402b	R22/R125/R290	Replaces R502
R403a (69S)	R22/R218/R290	Replaces R502
R403b (69L)	R22/R218/R290	Replaces R502
R408a	R22/R143a/R125	Replaces R502

HFC chlorine free blends (long term alternatives)

R404a	R143a/R125/R134a	Replaces R502 (R22)
R507	R143a/R125	Replaces R502 (R22)
R407a	R32/R125/R134a	Replaces R502 (R22)
R407b	R32/R125/R134a	Replaces R502 (R22)
R407c	R32/R125/R134a	Replaces R22

Other HFC alternatives

	R152a	Replaces R12
	R125	Replaces R502 (R13B1)
	R143	Replaces R502 (R13B1)
	R32	Replaces R22

Zeotropic blends

These are unlike azeotropic blends which behave as single substance refrigerants having a constant maximum and minimum boiling point. The zeotropic blends do not behave as a single substance during the evaporation and condensing processes. The phase change occurs with zeotropic refrigerants in a 'gliding' form over a particular temperature range. This temperature glide can vary and is mainly dependent upon the percentage and boiling points of the individual components of its composition. When a blend has less than 5K glide it is sometimes referred to as 'near azeotropic' or 'semi-azeotropic', R404 being an example of this.

Refrigerant blends

Many refrigerant blends suffer from what is known as 'glide'. The term boiling point is no longer applicable when dealing with these and instead the term 'bubble point' is used.

Briefly the bubble point can be regarded as the temperature at which vapour starts to form at a given pressure. If vaporization continues at the same pressure then the temperature at which the last drop of liquid refrigerant vaporizes is known as the 'dew point'. The dew point therefore is also the temperature at which condensation starts as a superheated vapour is cooled at a constant pressure.

With a pure refrigerant and with a true azeotrope (such as R502) there will not be any temperature changes in an evaporator as the liquid refrigerant vaporizes at constant pressure. With blends it becomes more complicated because the constituents have different volatilities. This means that they will vaporize at different temperatures. The blends do not boil at a constant temperature and pressure. Figures 116 to 118 give examples of the glide when plotted on a pH chart.

A plot for R12, R22 and R502 would be as shown in Figure 116.

A plot for the blends R404A and R407A would be as shown in Figure 117.

A plot for a refrigerant shows it to have a severe glide. An evaporator would be selected at a mean condition, as shown in Figure 118.

Figure 119 is a simple example of a plot on an enthalpy chart for a basic cycle with no liquid subcooling and no suction superheat. This illustrates bubble point, dew point, boiling range and glide. The cycle is based on 15 bar condensing pressure and 3 bar evaporator pressure.

Boiling range is the difference between dew point and bubble point at a given pressure. For 3 bar the boiling range $= -12 - (-19) = 7\,°C$.

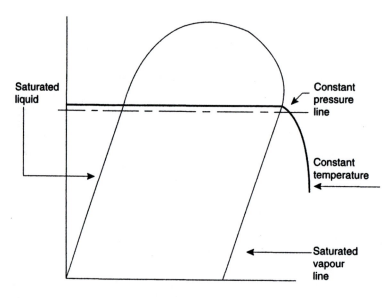

Figure 116 *A plot for R12, R22 and R502*

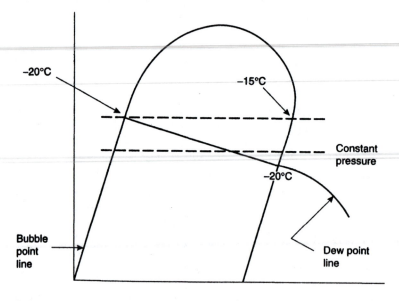

Figure 117 *A plot for blends R404A and R407A*

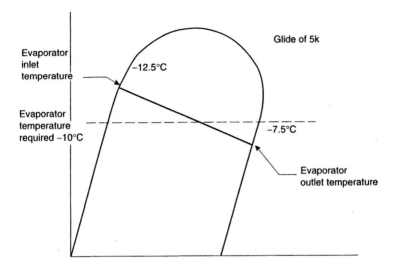

Figure 118 *A plot for a refrigerant with severe glide*

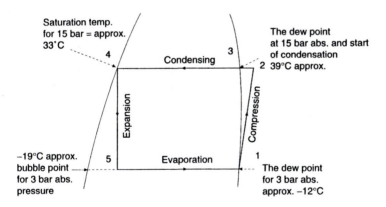

Figure 119 *A simple example of a plot on an enthalpy chart — basic cycle*

Glide is the difference in temperature in the evaporator:

At 5 the temperature is −17 °C approximately $\Big\}$ Glide = 5°
At 1 the temperature is −12 °C approximately

It will be seen from a study of the chart for R404A and R407A that when evaporating at a constant pressure the evaporating temperature rises and this can be as much as a 5K rise. The reverse occurs when condensing.

At constant pressure condensation can commence at 40 °C and finish at 37 °C.

Expansion valve setting

If an expansion valve is set for 5K superheat, the temperature at the thermal bulb position would be $(-7.5 + 5) = -2.5$ °C.

Manufacturers of coolers, units and expansion valves will be pleased to supply capacity ratings for their products upon request.

Klea 407A is suitable for operation at temperatures between -35 °C up to 45 °C with discharge temperatures that compare favourably with R502.

Klea 407B, when used in existing R502 systems within a temperature range of -40 to 40 °C and where discharge temperatures are critical, also compares favourably.

Klea 66 (407C) can be compared generally to R22 with lower discharge temperatures.

Changing the refrigerant charge

The actual task of replacing the refrigerant in an existing plant is not a simple one like replenishing a charge. Various service organizations no doubt have their own procedures. A few manufacturers have adopted the method known as retrofit procedure but this procedure is not common to all replacement refrigerants and it is imperative that the following points should be noted.

1 Check the existing system design and operating pressures to ascertain that they are compatible with the new refrigerant. The manufacturer of the equipment should be consulted if in doubt.
2 Change the compressor lubricating oil to that recommended according to the refrigerant used.
3 Flush the system through with the original refrigerant to reduce the percentage of oil contamination.
4 If necessary change shaft seals, filter driers, components containing elastomers and expansion valves to required types.
5 Adjust expansion valve superheat settings for optimum performance.
6 Reset pressure operated safety switches if necessary. If fusible plugs are installed check that the rating is satisfactory for operation with the replacement refrigerant.

A refrigeration system charged with a CFC or HCFC will probably use a mineral oil for lubrication. This type of oil may not be suitable for use with the replacement refrigerant. Before commencing any procedure it is advisable to consult the equipment manufacturer to determine suitability of the system components with the replacement refrigerant. It is also most important to determine the performance of the evaporator with a proposed replacement refrigerant.

A polyol-ester (PE) lubricating oil has been introduced which is generally a preferred choice. The PE lubricating oil may be used with some of the new refrigerants **but not all of them**.

For example, the stability of a system charged with HCFC 134a will possibly be reduced if the system contains chloride ions. HCFC 134a is an acceptable replacement for CFC 12 in a new installation which has not previously operated with a CFC. Before the system is commissioned the type of lubricating oil in the compressor should be checked. HCFC 134a **should not be used** in a system that has previously operated with CFC 12 until the system has been flushed extensively. Mineral oils normally used with CFC 12 are insoluble in HCFC 134a and **a suitable oil must be used**.

Replacement refrigerants may act as a solvent to elastomers or other materials used in the construction of shaft seals, filter driers and various types of valves. Gaskets between planed surfaces may also be suspect and once again it must be stressed that advice regarding system components should be sought to avoid unnecessary malfunctions.

Reclaiming the refrigerant

When the refrigerant in a system is to be removed and the system compressor is operational it is possible to use the compressor to recover the charge. Obviously, the arrangement of the service valves on the system will affect the exact procedure.

It is possible to pump down the system in the usual manner and decant the refrigerant into a cooled recovery cylinder. It may also be possible to use the recovery cylinder installed at the compressor discharge to act as both a condenser and receiver.

In cases where the compressor has failed and the system is to be charged a recovery unit could be used to reclaim the refrigerant. There are many types of portable recovery units available which can be connected to any system service valves. A recovery unit can handle refrigerants in either liquid or vapour form. It will remove a system charge to any suitable pre-set pressure. Most recovery unit capacities range from 50 to 100 kg/h and operate on single phase 220/240 V, 50 Hz electrical supply.

It is possible to recover refrigerant from a hermetically sealed system which has no service valves by using line tap valves following the recovery unit manufacturer's instructions.

Changing the refrigerant charge

As previously stated, exact procedures adopted by service agents may vary and the one given here is typical for replacing a refrigerant charge of CFC 12 with R134a.

The refrigerant in a system that requires replacement may be identified by the following methods:

(a)   Refrigerant type stamped on the compressor nameplate.
(b)   Refrigerant charge indicated on the expansion valve.
(c)   By the system standing pressure.

Procedure

1   Isolate the compressor after evacuating the vapour in the compressor crankcase.
2   Drain the old mineral oil from the crankcase.
3   Replace that oil with a suitable PE oil.
4   Open compressor service valves and operate the plant.
5   Repeat step 1.
6   Drain the PE oil from the crankcase.
7   Replace with fresh PE oil.
8   Open compressor valves and operate the plant.
9   Repeat step 1 again.
10   Drain the oil and carry out a contamination check. There must be less than 1% contamination by the original mineral oil. (This test requires a special oil test kit and is easily applied.)
11   If less than 1% residual mineral oil, recover the CFC 12 refrigerant from the system.
12   Change filter drier, expansion valve and any other component necessary.
13   Evacuate the system and replace the PE oil once more.
14   Charge the system with R134a refrigerant.

Note: After step 10, if the oil contamination is still in excess of 1%, steps 1, 2, 3 and 4 must be repeated, after which another oil test must be carried out.

In reality R134a should not be charged until the oil contamination is less than 1%.

The chlorine free refrigerant R134a was selected as an example because it is available in sufficient quantities to replace R12. Another reason for its selection was because there are probably a greater number of small R12 systems in the commercial field than the larger systems using other refrigerants. It can be used as an alternative for most R12 applications.

Compressor lubricating oils

The traditional mineral and synthetic oils are not miscible (soluble) with R134a. Oils which are not miscible can become entrained in heat exchangers in undesirable amounts to prevent adequate heat transfer, thus reducing the system performance.

The new lubricants developed, polyol-ester (PE) and polyalkene-glycol are miscible. These oils have similar characteristics to the traditional oils. They are more hygroscopic, dependent upon the solubility of the refrigerant. This means that the oils readily absorb moisture from the atmosphere.

Special care must be exercised during service, storage, charging, during dehydration and evacuation to avoid chemical reaction in systems such as copper plating. PAG oils tend to be more critical and are mainly used where high solubility is required. Ester oils are preferred by the industry in the main and information to date regarding their usage has proved satisfactory when the moisture content in the oil does not exceed 100 ppm. Experience has determined that systems should be well dehydrated and evacuated and relatively large drier capacities should be provided.

Other oils available and supplied by Castrol are alkylene benzene 2283 and 2284 which supplement the polyol-ester Icematic SW range.

Klea recommend EMKARATRL for use with Klea 407A, 407B and 407C application. This oil is fully compatible with these refrigerants.

Refrigerant reclaim

This refers to the process of removing refrigerant from a system to be passed through a reclaim unit using large capacity filter driers which are capable of retaining moisture and acid content. The reclaimed refrigerant is then discharged by the reclaim unit into refrigerant cylinders for re-use. Alternatively the reclaimed refrigerant can be despatched for processing by specialists.

Refrigerant recovery

By way of an example assume that a condenser was found to have developed a slight refrigerant leakage. The leak being on the high pressure side of the system rules out the possibility of any moisture contamination. Under these circumstances the refrigerant could be recovered by using a recovery unit, discharged into a service cylinder or cylinders if the charge is large. When repairs to the condenser have been completed the condenser should be pressure tested and evacuated. The recovered refrigerant can then be recharged into the system and the operating charge 'topped up' with the same refrigerant if necessary.

Due to the boiling points of various substances in refrigerant blends the actual recharging or initial charging of systems should be in liquid form. If a system is vapour charged there is a danger of drawing off from the refrigerant cylinder a greater quantity of one of the substances constituting the blend. This would undoubtedly alter the percentage of the mixture.

When reclaiming or recovering refrigerant from a commercial plant, whether the system compressor is operational or not, most of the refrigerant will be in the receiver and condenser in liquid form. The refrigerant is best removed from a system in liquid form using a reclaim unit. If it is possible to 'pump down' a system this task is much easier.

Obviously the process will be carried more rapidly and more efficiently if the refrigerant is drawn from both the high and low pressure sides of the system. The valve plate assembly of the compressor, expansion valve or the capillary will resist refrigerant passing through them. Inevitably some refrigerant will remain entrained in the oil which will settle in various parts of the system, for example the evaporator and compressor. If a compressor is fitted with a crankcase heater it can be actuated for a short period to raise the temperature of the crankcase oil which will then release the refrigerant from the oil. Defrost heaters on an evaporator can be employed in a similar manner.

Caution: Do not overheat the compressor or evaporator. The use of flame producing devices is not recommended for this purpose.

During a reclaim/recovery operation it is essential that sufficient cylinder capacity is available and double valve cylinders are desirable. Where possible cylinders should be cooled. **Cylinders should never be overfilled**.

It is an advantage to install an oil container between the recovery/reclaim unit and the liquid receiver of the system.

Refrigerant hoses

Most manufacturers of pressure hoses supply them complete with Shraeder connectors. These can be removed to further reduce any resistance to the flow of refrigerant during the reclaim/recovery operation.

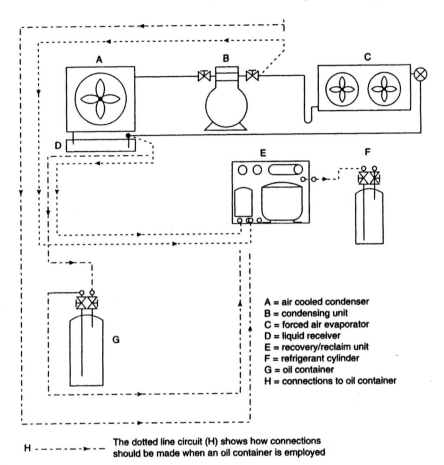

A = air cooled condenser
B = condensing unit
C = forced air evaporator
D = liquid receiver
E = recovery/reclaim unit
F = refrigerant cylinder
G = oil container
H = connections to oil container

H ------►-- The dotted line circuit (H) shows how connections
should be made when an oil container is employed

Figure 120 *Commercial system hose connection arrangement*

Domestic and commercial sealed hermetic systems

Domestic refrigerators and freezers do not normally have liquid receivers but access to both high and low pressure sides of the system can be achieved by using line tap valves. Some small commercial sealed systems may well fall into this category.

Other commercial sealed systems will have liquid shut-off valves at the receiver and suction service valves at the compressor.

Reference to the reclaim/recovery unit arrangement will show how hose connections can be made.

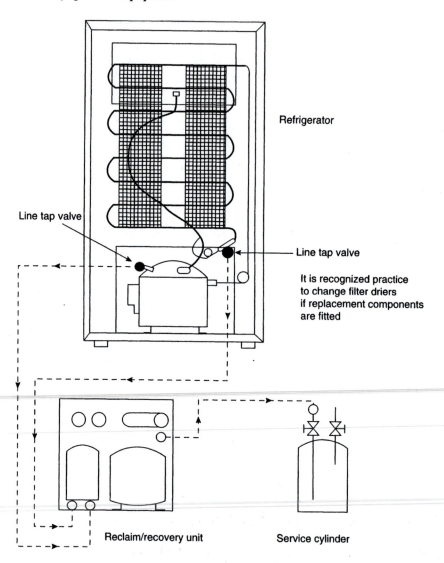

Refrigerator

Line tap valve

Line tap valve

It is recognized practice
to change filter driers
if replacement components
are fitted

Reclaim/recovery unit

Service cylinder

Figure 121 *Domestic refrigerator system hose connection arrangement*

Table 14   *Pressure—temperature relationship*

Temp. (°C)	Estimated pressure (bar) at 1 atm								
	R134a	R407A	R407B	R407C	R32	R125	R402A	R404A	R508
−50	0.2	—	—	—	1.1	0.9	—	—	5.6
−40	0.5	—	1.1	—	1.7	1.4	1.5	1.3	8.4
−30	0.8	1.5	1.8	1.4	2.7	2.2	2.4	2.0	11.8
−20	1.3	2.1	2.3	2.0	4.0	3.3	3.5	3.0	15.1
−10	2.0	3.2	3.4	3.1	5.8	4.8	4.8	4.1	—
0	2.9	5.0	5.1	4.1	8.1	6.7	6.5	6.0	—
10	4.1	8.0	7.1	6.0	11.0	9.0	9.1	8.0	—
20	5.6	9.1	10.0	8.1	14.7	12.0	10.2	10.1	—
25	6.6	10.1	12.2	10.5	16.8	13.7	13.5	12.0	—
30	7.6	12.8	13.8	11.8	19.2	15.6	15.0	14.5	—
40	10.1	15.1	18.0	15.0	24.8	20.0	19.7	15.3	—
50	13.1	21.8	21.4	20.0	31.4	25.3	25.2	20.1	—
60	16.7	—	—	—	39.4	—	—	—	—
70	21.0	—	—	—	48.8	—	—	—	—
80	26.2	—	—	—	—	—	—	—	—
Bubble point at 1 atm. (°C)	—	−45.5	−47.3	−44.0	—	—	−48.9	−46.9	−85.6
Bubble point pressure (bar)	—	12.6	13.3	11.9	—	—	13.4	12.6	8.2
Estimated critical temp. (°C)	101.0	83	76	86	78.3	66.3	81.0	72.0	23
Estimated critical pressure (bar)	40.5	—	—	—	58.1	36.4	42.7	—	40.6
Boiling point (°C)	−26.2	—	—	—	−51.7	−48.2	—	—	—

Note: The term 'bubble point' is used with refrigerants composed of a number of fluids. Each fluid will have a boiling point and will vaporize at a different temperature and pressure. The bubble point temperature therefore is the temperature at which the mixture or blend of the refrigerant starts to vaporize.

Table 15 *Pressure–temperature relationships for FREON refrigerants*

Temp. (°C)	Vapour pressure (kPa)			
	FREON			
	CFC 12	HCFC 22	R-500	R-502
−100				3.2
−95				4.9
−90				7.2
−85				10.4
−80			10.5	14.6
−75		8.8	14.8	9.2	20.3
−70		12.3	20.6	12.8	27.6
−65		16.8	28.1	17.6	36.9
−60		22.6	37.7	23.7	48.7
−55		29.9	49.7	31.3	63.4
−50		39.0	64.7	41.0	81.4
−45		50.3	83.1	52.9	103.3
−40		64.0	105.4	67.4	129.6
−35		80.5	132.2	85.0	161.0
−30	9.2	100.1	164.0	105.9	197.9
−25	12.2	123.3	201.6	130.7	241.0
−20	15.8	150.5	245.4	159.8	291.0
−15	20.3	182.1	296.3	193.6	348.6
−10	25.8	218.5	354.9	232.8	414.3
−5	32.5	260.3	421.9	277.8	488.9
0	40.4	308.9	498.2	329.1	573.1
5	49.9	361.8	584.4	387.3	667.6
10	61.0	422.6	681.4	453.0	773.1
15	74.1	490.6	789.9	526.8	890.2
20	89.2	566.6	910.9	609.2	1019.7
25	106.6	650.9	1045.1	701.0	1162.3
30	126.6	744.2	1193.4	802.6	1318.9
35	149.4	847.0	1356.6	914.9	1490.1
40	175.3	960.0	1535.7	1038.3	1677.0
45	204.4	1083.6	1731.4	1173.7	1880.3
50	237.2	1218.4	1944.8	1321.6	2101.3
55	273.7	1356.1	2176.7	1482.8	2341.1
60	314.5	1524.2	2428.0	1657.9	2601.4
65	359.6	1696.3	2699.6	1847.6	2884.0
70	409.4	1882.0	2992.3	2052.7	3191.8
75	464.3	2082.0	3307.2	2273.9	3528.5
80	524.5	2296.7	3645.7	2511.9	3900.4
85	590.3	2527.0	4007.3	2767.5	
90	662.0	2773.9	4393.8	3041.7	
95	739.9	3036.6	4806.1	3334.5	
100	824.4	3316.8		3647.1	
105	915.8	3615.0		3980.4	
110	1014.4	3931.9			
115	1120.5				
120	1234.4				
125	1356.6				
130	1487.2				
135	1626.6				
140	1775.3				
145	1933.4				
150	2101.3				

Table 16 *Pressure–temperature relationships for SUVA refrigerants*

Temp. (°C)	Vapour pressure (kPa)				
	SUVA Centri-LP	SUVA Chill-LP	SUVA Freez-HP	SUVA Cold-MP	
		HCFC 124		HFC 134a	HFC 23
−100					31.6
−95			4.8		44.9
−90			8.2		62.5
−85			11.9		85.1
−80			16.9	2.5	113.9
−75			23.4	4.5	149.9
−70		2.9	32.0	8.3	194.3
−65		5.0	42.9	11.7	248.2
−60		8.0	56.6	16.3	313.1
−55		11.1	73.6	22.3	390.3
−50		15.1	94.6	29.9	481.2
−45		20.3	120.0	39.6	587.4
−40		26.8	150.6	51.7	710.3
−35	4.3	34.9	187.0	66.6	851.8
−30	6.7	44.9	229.9	84.8	1013.2
−25	9.2	57.1	280.2	106.7	1196.5
−20	12.1	71.8	338.5	133.0	1403.3
−15	15.8	89.4	405.9	164.1	1635.3
−10	20.4	110.2	483.1	200.6	1894.4
−5	26.0	134.7	571.1	243.2	2182.4
0	32.8	163.2	670.9	292.5	2501.3
5	41.0	196.2	783.3	349.2	2852.7
10	50.7	234.1	909.5	414.0	3238.6
15	62.2	277.5	1050.5	487.7	3661.7
20	75.7	326.9	1207.3	571.0	4122.3
25	91.4	382.6	1381.0	664.7	4623.6
30	109.6	445.3	1572.7	769.7	
35	130.6	515.5	1783.6	886.7	
40	154.5	593.8	2014.8	1016.6	
45	181.7	680.6	2267.5	1160.4	
50	212.5	776.7	2543.3	1318.8	
55	247.2	882.5	2842.1	1492.9	
60	286.1	998.7	3165.9	1683.6	
65	329.4	1125.9	3516.0	1891.7	
70	377.7	1264.7		2118.3	
75	431.1	1415.7		2364.4	
80	490.0	1579.6		2630.9	
85	554.7	1757.0		2919.6	
90	625.8	1984.4		3229.8	
95	703.3	2154.6		3563.8	
100	787.9	2376.2		3922.4	
105	879.7	2614.4			
110	979.3	2868.5			
115	1086.9	3140.1			
120	1203.0	3429.7			
125	1327.9				
130	1462.0				
135	1605.7				
140	1759.4				
145	1923.5				
150	2098.3				

To convert kPa to bar divide by 100. kPa ÷ 100 = bar abs.

Appendix A Fundamental principles of air conditioning

To achieve effective air conditioning of a room or space requires simultaneous control of the following:

Temperature
Humidity
Air cleanliness
Air motion.

Environmental conditions within a room or space must be controlled to provide a comfortable atmosphere for human occupation. Certain manufacturing processes also require a high degree of accuracy in air quality. This could embrace the drug industry, computers, hospitals and communications.

Air circulating within a room or space conveys heat, moisture, smoke, fumes, dust, dirt, odours, pollen and noise. These all have an effect upon human comfort as well as industrial processes. Therefore air conditioning can be classified as:

(a) Comfort air conditioning.
(b) Industrial air conditioning.

Consideration must be given to climatic conditions and seasonal changes. This means that an air conditioning system must function efficiently and economically throughout the year in order to provide:

(a) Summer air conditioning.
(b) Winter air conditioning.

It could be said that summer air conditioning serves to reduce temperature and to dehumidify whilst winter operation serves mainly to increase both temperature and humidity.

Temperature

The most important is heat and the amount of heat conveyed in the air undoubtedly has a direct effect on human health and comfort. It will also affect the growth of bacteria and moulds. A warm dry atmosphere can affect a person's respiration. It can be the cause of a dry throat and irritation of eyes and ears. An increase or decrease in temperature will cause expansion and contraction of metals thus creating problems in the production of precision engineered components. Complete comfort cannot be obtained by heat alone and the amount of moisture carried in the air is equally important. Warm air will hold more moisture than cold air and the condition of the air will constantly vary, day by day, season by season and of course by the location.

An air conditioning plant will reduce the sensible temperature of a space to a greater extent when the moisture content of the air is low than when it is high. Less refrigerant will be vaporized in the cooling system when condensing water vapour. Since moisture in the air is in a gaseous form and at a temperature below its boiling point it is referred to as water vapour.

The condensing of water vapour in the air requires the removal of heat from the water vapour and it is this 'latent heat' which vaporizes the refrigerant in the cooling system. The greater the amount of moisture or water vapour in the air the greater will be the 'latent load' and less refrigerant will be available for reducing the sensible temperature of the conditioned space.

In air conditioning for cooling or heating, reference to the quantity of moisture or water vapour in the air is defined as percentage saturation or relative humidity.

Composition of the atmosphere

With any form of air conditioning it is essential that something is known of the earth's atmosphere. Air is an invisible, odourless and tasteless mixture of gases which surround the earth. It extends approximately 400 miles or 650 kilometres above the earth and is divided into several layers. The layer closest to earth is known as the lower atmosphere which extends from sea level to approximately 30 000 feet (9150 metres). The next layer, the troposphere, extends from 30 000 feet to 50 000 feet (15 250 metres). The layer above the troposphere is known as the stratosphere and extends up to 200 miles or 320 kilometres. The layer beyond that is known as the ionosphere.

Atmospheric air comprises of a mixture of oxygen, nitrogen, carbon dioxide, hydrogen, sulphur dioxide, water vapour and minute percentages of various rare gases. Table A1 gives the proportions of dry gases in the air below an altitude of 15.5 miles or 25 kilometres.

The carbon dioxide content will vary due to the actions of living organisms and the burning of fossil fuels. Water vapour is also a variable factor since it is affected by the latitude and seasons of the year. The amount of water vapour associated with air temperature is greater during the summer months in low latitudes except for the arid tropical desert areas.

While oxygen in the air sustains life the nitrogen dilutes active oxygen to prevent oxidization of body tissues. Oxygen readily mixes with other substances. When fuel such as wood, coal or oil are burned the oxygen in the air mixes with the carbon and hydrogen to form carbon dioxide and water.

Oxygen in the atmosphere is replenished by growing plant life. The roots absorb moisture from the soil and oxygen from the moisture is released from the plant foliage.

Nitrogen, oxygen, argon and carbon dioxide are the four main gases, which constitute 98.98% of the air by volume. This varies very little up to a height of 50 miles or 80 kilometres. Water vapour is contained in the air up to a height of 6 miles or 10 kilometres and is virtually non-existent two or three kilometres above that.

Table A1

Component	Chemical symbol	% volume (dry air)
Nitrogen	N_2	78.08
Oxygen	O_2	20.944
Carbon dioxide	CO_2	0.03 (variable)
Hydrogen	H	0.00005
Helium	He	0.0005
Neon	Ne	0.0018
Ozone	O_3	0.00006
Argon	Ar	0.93
Krypton	Kr	Trace
Xenon	Xe	Trace
Methane	Me	Trace
Sulphur dioxide	SO_2	Trace

Physical properties of air

Air has weight, density, specific heat, heat conductivity and in motion it has momentum and inertia. It can hold substances in suspension, in the form of solids and solution.

Air pressure on the earth's surface is due to the weight of the air above the earth. This pressure decreases with increase of altitude because of the reduction of weight of air above it. At sea level atmospheric pressure against the earth is 14.7 pounds per square inch or 6.6 kilograms per square centimetre. It is sometimes expressed as 101 kPa and in meteorology as 1010 mb.

The air, having weight, requires energy to move it and once in motion it has energy of its own and this is known as kinetic energy. By example, the weight of moving air can turn the sails of a windmill which in turn converts the kinetic energy of the moving air into mechanical energy.

Velocity

This is measured in either feet per second or metres per second and increasing the velocity of the air decreases the air pressure. One example of this can be related to the low pressure of a tornado which can cause structural damage to buildings. When the pressure on the outside of a building is reduced quickly the pressure inside creates an outward force of moving air.

Dust particles carried by winds may be held in suspension within a mass of moving air for long periods. At certain times of the year dust from the Sahara desert is frequently precipitated in Europe. Many vehicle owners can give evidence of this when they have parked a vehicle outside overnight and after a heavy dew, they discoverd that the vehicles were covered in a fine film of dust.

The density of the air will vary with atmospheric pressure and humidity; 1 kg of air at sea level will occupy a space of approximately 0.84 m^3 with a density of 0.002 kg/m^3.

Basic psychrometry and definitions

As previously stated, atmospheric air is comprised of a mixture of gases and psychrometry is the name given to the study of such properties and the behaviour of the mixture. To determine the state of air a psychrometric chart (Figure A2) has been plotted using the basic properties of mass and energy, moisture and enthalpy as co-ordinates.

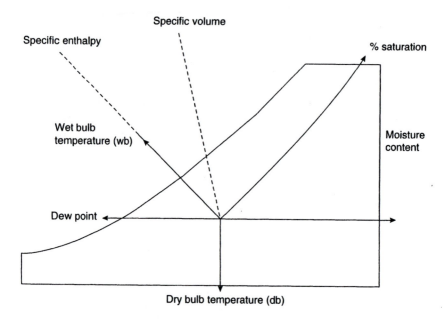

Figure A1

The psychrometric chart is a graph of the properties, temperature, humidity, volume etc. of atmospheric air. It is used to show how the properties will vary as the amount of water vapour in the air increases or decreases. If any two properties of the moist air are known all of the remaining properties can be determined from the chart. The simple diagram of the chart (Figure A1) provides further indication of the lines to trace when making a plot.

- **Dry bulb temperature**: Signified as °C db, this will record the temperature of a substance of moist air. It is measured by a thermometer with a dry sensing element.
- **Wet bulb temperature**: Signified as °C wb, this will show the temperature of the air measured by a thermometer with a sensing element permanently wetted by a wick immersed in a small container. When the thermometer is placed in a fast moving air stream evaporation of the water from the wick will lower the temperature of the wick. Heat will then flow from the thermometer to the cooler wet wick to record a lower temperature. If the air passing over the wet bulb thermometer is dry, evaporation from the moisture laden wick will be faster than if the air is moist. To obtain

both of these temperatures simultaneously an instrument called a sling psychrometer may be used.

- **Humidity**: This is the term used to describe the presence of water vapour in the air and the amount of water vapour it will hold will depend upon the temperature of the air. A person occupying a space where the temperature and humidity is high will feel warm and uncomfortable, and will possibly perspire because the rate at which perspiration will evaporate will be slow. Moist air prevents rapid evaporation. In an environment where humidity is lower the perspiration will evaporate much faster and as a result the person will feel cooler and more comfortable.
- **Relative humidity**: Signified as % rh, this is the term used to express the amount of water vapour in a sample of air as compared with the amount it would hold if it was saturated at the temperature of the sample. Relative humidity is given as a percentage of air saturated. It is a ratio of the moisture content of the air at a given temperature to the moisture of the saturated air at the same temperature.

    To put it more simply, saturated air is the air which contains all the moisture it can hold at any given temperature without precipitation. Relative humidity can be determined by comparing the actual moisture content with that at saturation, e.g.:

$$\text{Relative humidity} = \frac{\text{Actual moisture content}}{\text{Saturation moisture content}} \times 100$$

- **Moisture content**: Signified as kg/kg mc, this is the term used to express the actual weight of water vapour present per kilogram of air.
- **Specific enthalpy**: Signified as kJ/kg, this represents the total heat content associated with a unit mass of air at a temperature above the datum of °C.
- **Specific volume**: Signified as m^3/kg, this represents the volume occupied by 1 kg of atmospheric air at a given temperature.
- **Dew point**: Signified as °C dp, this term defines the temperature to which saturated air can be cooled without precipitation. It can also be defined as the temperature at which the vapour pressure of a sample of moist air is equal to the saturated vapour pressure. Any reduction of temperature of air at its dew point would result in precipitation. Precipitation is sometimes referred to as 'raining out' and this would be seen as condensation.

The psychrometric chart

To a person unfamiliar with it the psychrometric chart (shown in Figure A2) would appear to be a mass of lines. It is in fact three triangular graphs each

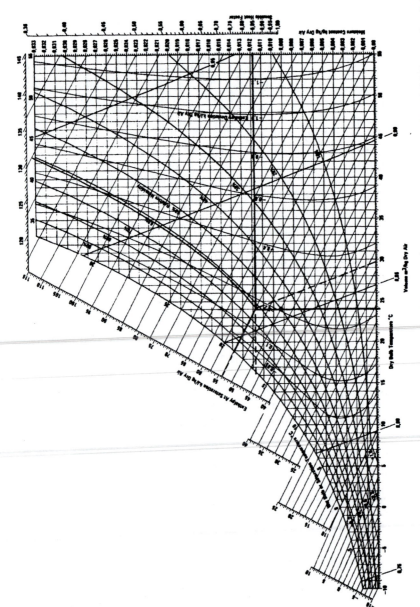

Figure A2 Psychrometric chart. Normal temperatures, SI metric units, barometric pressure 101 325 kPa, sea level. Below 0°C properties and enthalpy deviation lines are for ice. © Carrier Corporation

superimposed over the other. The first represents the dry bulb temperature, the second the wet bulb temperature, the third the dew point temperature. Using a sling psychrometer to determine the dry and wet bulb temperatures provides the two factors required. Once these are established it is possible to find the dew point temperature, moisture content, specific enthalpy and specific volume. The lines to follow are given on Figure A1. The purpose of the sling psychrometer is to obtain the temperature difference between the dry and wet bulb. This temperature difference is called 'wet bulb depression' and the wb temperature will always be the lowest.

Reading the chart

Assume that the db and wb temperatures of a sample of air are recorded as being 24 °C and 19 °C respectively. If the db line is traced until it intersects with the wb line it will be seen that this is at a point just above half way up the 24 °C line. Tracing the curve upwards to the right of this point will show that the air sample is 61% saturated.

A horizontal trace to the left gives a dp temperature of 16.5 °C. The dotted lines above the intersection indicate both specific enthalpy at 53.6 kJ/kg and specific volume of 0.86 m^3/kg. These values have been plotted on the psychrometric chart as an example.

There are many different designs of sling psychrometers available but the basic design remains the same and when using this instrument the steps below must be observed.

1   Ensure that the wick or gauze is adequately wetted and, if the instrument is designed with a reservoir, that this is filled with water. Distilled water should always be used.
2   Whirl the instrument in the air sample at a rate between 150 and 250 rpm for a period of 15 to 25 seconds.
3   Record the db and wb temperatures, repeating step 2 three or four times and the lowest wb temperature should be used as the recorded temperature.
4   Make a plot on the psychrometric chart to determine the relative humidity.

There are other instruments available for measuring humidity such as a wall type hygrometer or electric hygrometer.

Just as temperature affects the comfort of a person so does relative humidity. Low atmospheric humidity will increase the amount of electrostatic energy in the air. An indication of this energy existing may be evident when a person touches a metal object which is 'earthed' and a slight electric shock is felt. A person removing clothing in a dark room may often observe minute blue sparks as the articles are removed. Women may find that hair styles become unmanageable. Joints in woodwork tend to contract and become loose, cracks may appear in woodwork over a period of time.

Humidity control

This is an important factor in air conditioning and various controls are employed to keep the humidity of a space at a satisfactory level. One such control is the humidistat which operates in the winter to ensure that the system will add moisture to the air (humidify). During the summer it will operate to remove moisture from the air (dehumidify) to maintain a fairly constant level of humidity.

This control operates electrically to open or close solenoid valves, dampers or air by-passes to regulate the flow of air over the system cooling coils. The sensing element of this type of control may be of synthetic fibre or even human hair which is very sensitive to the moisture content of the air. The sensing element will contract or stretch to activate a switch mechanism according to prevailing humidity conditions. The electrical supply is then connected or disconnected to the solenoid circuits. The length of the sensing element increases with an increase of humidity and decreases as the moisture content of the air is reduced. When installed in an electrical circuit of an air conditioning system it will normally override the thermostatic control when humidity is high.

Air motion

Discomfort to occupants of a space can be attributed to improper air movement even when the temperature and humidity are correctly maintained. When cool air is circulated to a warm atmosphere heat from the atmosphere will flow from that space, and from the objects and occupants within it. As heat flows to the cooler air evaporation will increase, thus creating the cooling effect. Sometimes a person exposed to atmospheric conditions will feel much colder than the actual temperature indicates. This could be due to both relative humidity and wind velocity. Weather forecasters refer to this as the 'wind chill factor'. Air movement is essential to suppy an adequate quantity of fresh air to a controlled space. If the air movement is too fast it will cause draughts and occupants within the space could feel some discomfort. If the air movement is too slow then the atmosphere within the space will become stale and lack a degree of oxygen—sufficient to cause drowsiness.

Air movement

This is defined as the distance travelled per unit of time. If the air velocity is multiplied by the cross sectional area of a duct the volume of air flowing through it can be calculated and this is expressed in cubic metres per second.

Measurement of air velocity

Different methods and various instruments are used to perform this task. Installation and service engineers specializing in air conditioning should have a sound knowledge of their construction and operation. These instruments are:

- the rotating vane anemometer,
- the pitot tube (velocity pressure),
- the hot wire anemometer, and
- the velometer.

The rotating vane anemometer, velometer and pitot tube are not accurate at extremely low velocities. The choice of instrument will largely depend upon site conditions, if the air is flowing into or from a duct, inside a duct or to determine the air movement within a controlled space.

The rotating vane and hot wire anemometers should be located in the air stream at right-angles to the flow. They should be level and allowed to remain in the air stream for at least one minute until a constant flow of air passes over the instrument and a true reading is attained. The rotating vane type is by far the simplest instrument to use. The rotating speed of the air is converted into measurement of distance by a gear mechanism. This is then indicated on dial gauges which are an integral part of the design.

The hot wire type relies upon the cooling effect of the air flow passing over an electrically heated wire located inside a probe. This instrument can be used in duct air streams, open spaces and at the inlet or outlet of ductwork grilles and diffusers.

The velometer is used in a similar manner to the rotating vane anemometer to give direct readings of velocities. Although suitable for ducts, air flow, inlet and outlet of system ductwork, it cannot be used for recording low velocity air movement in a conditioned space.

When any of these instruments are used several readings should be taken at each location, at different points of the duct system. An average figure can then be taken and recorded.

When measurements are made at duct inlets and outlets the grille area should be divided into 150 mm square sectors. The readings should then be taken from the centre of each sector. With a square grille this is simple but with a circular grille a number of readings should be taken from around the duct grille in equal circular areas. An average figure can then be used. The volume of air can be recorded thus:

Volume = Velocity of air × Free area of grille

Velocity pressure (pitot tube)

Obviously pressure will exist within the ductwork of an air conditioning system and to measure this pressure a manometer is used together with a pitot tube. The total pressure within a duct is the sum of two pressures.

1   Static pressure.
2   Velocity pressure.

Static pressure develops in all directions inside a duct because the duct surface will resist air flow due to friction. This resistance must be overcome so that air flow along the duct can take place. Static pressure inside the duct can be positive or negative, relative to barometric pressure prevailing. The negative pressure will be evident in the inlet duct as the air will be under suction from the fan(s). The air discharged by the fan(s) will create a positive pressure. Velocity pressure is produced by the air movement and in the direction of the air flow.

To measure the pressures inside the ductwork system a manometer and pitot tube are employed.

The pitot tube actually comprises two tubes, one located inside the other. The outer tube has perforations along the side of it to allow air to enter. This tube measures the static pressure. The inner tube measures the total pressure and from these two pressures the velocity pressure can be determined. When the two pressures are connected to a manometer the difference in pressures is the velocity pressure. An example of a pitot tube connected to an inclined manometer is shown in Figure A3.

Comfort conditions

Since air conditioning in the past was generally provided for the comfort of occupants in homes, offices and other areas of work the term comfort conditions was applied.

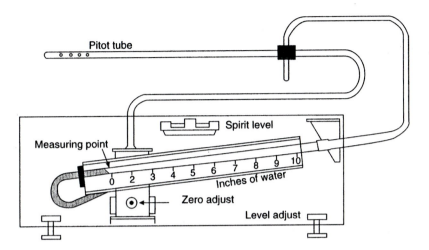

Figure A3 *Pitot tube and inclined manometer*

It is customary for inhabitants of hot countries to wear loose fitting clothing to minimize evaporation of perspiration. The relative humidity plays a large part in producing clammy or cool conditions.

Air movement also is a distinct contributing factor because the air moving around a person aids the evaporation process. Experience has proved that air moving at a rate of 0.08 to 0.12 m/sec may result in providing a stagnant atmosphere. A lower db temperature would then be required to give comfort conditions, particularly when the air supply is heated prior to it being circulated to a conditioned space. Cooling air movement in excess of 0.12 m/sec may produce a slightly higher db temperature but increasing the rate of air movement can create draughts.

Another factor to be considered is the building design; large window areas will mean that radiated heat from the sunlight and reflecting surfaces will add to the space temperature. The distance the window area is from where the occupants are located is also a factor. The window areas and radiant surfaces will possibly be at a higher temperature than the air in the conditioned space. To compensate for this radiated heat the dry bulb temperature will need to be lowered in order to maintain the desired comfort conditions.

Effective temperature

The db temperature and relative humidity combined produces the effective temperature. This is a relationship where a higher db temperature with a lower

Table A2

Dry bulb temperature (°C)	Relative humidity (%)	Effective temperature (°C)
20	100	20
22	60	20
25	15	20

relative humidity is considered acceptable to the majority of the occupants of a conditioned space. A study of the comfort chart (Figure A4) shows that the central zone gives effective temperatures when plotted as a combination of the db temperature and relative humidity. Each effective temperature on the chart starts at the same value as the db temperature on the 100% relative humidity line.

Range conditions

At an approximate relative humidity range between 70% and 40% the comfort condition on an effective temperature line would be considered the same. With a relative humidity above 70% body cooling is not usually enough to provide comfort conditions. Below 40% relative humidity, occupants of a space could experience a dryness of the mouth, nose and throat.

It can be seen that a range of conditions can be satisfactory for human comfort without it being necessary to provide a set condition for both summer and winter. The summer effective temperature at the top of the comfort zone is 21 °C at 50% relative humidity with a db temperature of 23.5 °C. At the bottom of the comfort zone the effective temperature is 18 °C at 50% relative humidity and an ideal db temperature of 20 °C. This might be suitable for winter conditions but a slightly lower db temperature could be acceptable.

One other factor must be considered and this is the amount of activity by the occupants of the conditioned space. A lower temperature is often desirable when there is a high degree of activity, a higher temperature being necessary when occupants are at rest.

It must be realized that occupants moving to a conditioned area from a warmer humid area or extremely high ambient temperature can be put to some discomfort due to the temperature difference. The sudden change in temperature can produce what is known as 'thermal shock'. This is the result of a person's body not having sufficient time to acclimatize to the change

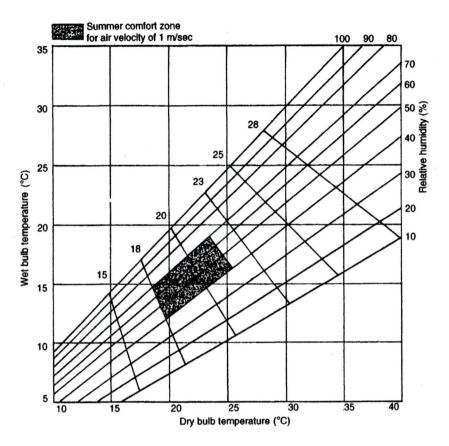

Figure A4 *Comfort chart*

of conditions. A temperature difference of not more than 12 °C in summer and 16 °C in winter is generally recommended where frequent movement of personnel is involved. It has been scientifically proven that personnel function far more efficiently within a conditioned environment. Persons sleeping, eating and indulging in leisure activities also benefit when taking place in comfort zone conditions.

To summarize, the function of any air conditioning system is to maintain desirable specific conditions within a room or space. This means that it has to supply heat when required, remove it when there is a heat gain and either induce or remove moisture from the air. In addition it should be capable of removing dirt, dust particles and pollen from the air.

It is not intended to deal with the countless number of designs and systems in this publication. Not all systems provide all of the factors mentioned. A diagram of a system which will do so (Figure A5) is given for reference and component identification. Figure A5 depicts a complete system and is typical of many in general use today.

As the title of this handbook implies, it deals with service and installation of refrigeration equipment which is a major part of any air conditioning system. Malfunction or failure of the refrigeration plant in any system will occur and these have been dealt with in previous chapters. However, a service or maintenance engineer, or personnel engaged upon the installion of air conditioning

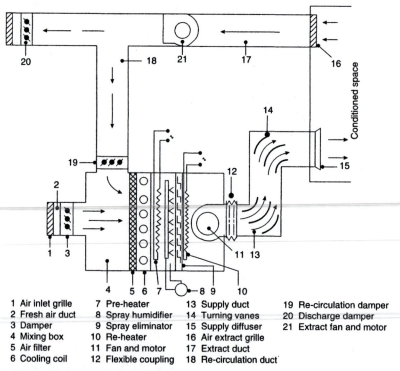

1 Air inlet grille	7 Pre-heater	13 Supply duct	19 Re-circulation damper
2 Fresh air duct	8 Spray humidifier	14 Turning vanes	20 Discharge damper
3 Damper	9 Spray eliminator	15 Supply diffuser	21 Extract fan and motor
4 Mixing box	10 Re-heater	16 Air extract grille	
5 Air filter	11 Fan and motor	17 Extract duct	
6 Cooling coil	12 Flexible coupling	18 Re-circulation duct	

Operating sequence: Fresh air is drawn into the mixing box to pass through a filter before entering the cooling coil. After leaving the cooling coil the air is then pre-heated to be humidified. The humidified air is then re-heated to be circulated to the conditioned space. The spray eliminator is to prevent the water spray passing to the supply fan chamber. Turning vanes are installed to minimize eddy currents of air in the ductwork where a change of direction takes place. Some of the extracted air from the conditioned space is routed back to the mixing box to be re-processed and re-circulated with the fresh air.

Figure A5   *Example of a complete air conditioning circulation system*

systems will at some time be confronted with noise problems which do not originate from the refrigeration plant.

Maintenance – air flow and noise problems

In addition to the cooling, heating and temperature conditions consideration has to be given to the velocity and circulation of air within a space. The operation of fans and the condition of system ductwork is important. There are three types of fans employed with air conditioning systems: axial flow, radial flow and centrifugal.

- **Axial**: Air enters and leaves this type of fan axially, giving a straight-through air flow. It is inherently noisy and seldom used for comfort air conditioning. It is acceptable in some cases for industrial installations where the noise factor is not so important.
- **Radial**: This type of fan can deliver a straight-through air flow being mounted within the ductwork with the perimeter fan blades in the same direction as the air stream. It circulates air in much the same manner as a paddle wheel and is reasonably quiet.
- **Centrifugal**: This type is most favoured for low-noise value and its ability to perform extremely well against ductwork resistance pressures.

Ductwork

There are two types of ductwork used: round and square or rectangular. The round duct, although it is made up of less material, is easy to manufacture and cheaper to produce, does not take preference over the other type. This in spite of the fact that it can handle a greater volume of air with less ductwork area. Also, it does not present as much resistance to air flow as the square or rectangular type.

The square or rectangular designs are selected generally because modern building designs enable it to be easily concealed in available spaces to provide a more attractive appearance.

Any ductwork will create resistance to air flow just as for water flowing through a pipe. Both the duct and pipe will resist the flow.

Air passing along ductwork is subject to friction loss and this has a direct relationship to the length and perimeter of the ductwork, but that is a question of design. The number of direction changes, loss through cooling coils, heater grids, filters and the condition of the duct interior surface area are all contributing factors. It is therefore imperative that all ductwork is kept clear of obstructions, that coils are cleaned and that filters are changed or cleaned regularly.

After installation and prior to commissioning of new plant, during routine maintenance and after some service operations, the following checks should be made. This will minimize noise and vibrations from an installation.

1   Obstructions in ductwork upstream and downstream of fans must be removed. High-velocity air striking sharp metal objects inside ducts and metal edges on grilles are a major source of noise. A low-pitched rumbling sound is usually caused by fan or motor noise being transmitted along ductwork. A popping sound evident when the unit starts or stops is generally the result of expansion and contraction of the ductwork as it warms and cools. Should this be at an unacceptable level it can be eliminated by insulating the ductwork.

2   Some ductwork incorporates flexible inlet and outlet connections to the fan housings. These are for minimizing and isolating fan noise. They should be installed level with the duct openings and the flexible connection to the duct and fan housing must be taut. Any slackening of the connection will cause billowing with eventual fracture. It will also cause eddy currents in the air flow within the ductwork to affect the air flow and possibly create a noise factor.

3   Electrical motor cables inside the duct should be flexible, coiled and secured. Resilient motor mountings should be regularly inspected for wear, distortion or misalignment. If fan motors have a solid base, anti-vibration pads should also be inspected. Any noise from components making metallic contact will be amplified by the ductwork. All electric motors should be oiled before start-up and during routine maintenance.

4   The fans of some systems are belt driven from an electric motor and it could possibly pose another noise factor. Ensure that the belt tension is correct and that the drive pulleys are fitted as close to the fan and motor bearings as possible. This will reduce bearing load and it is a well known fact that misaligned pulleys account for a considerable number of unnecessary service calls. Fan and motor operation should always be checked again after the initial running period or when new belts are fitted (belt stretch and alignment).

5   It is an accepted practice to operate the air distribution fan(s) only for a considerable period after any installation or major routine check. This will assist in locating any source of noise or vibration generated by the distribution fan(s). By adopting this procedure any noise or vibration will not be associated with that generated by operation of the condensing unit or refrigeration system pipework. Isolating air flow fans and operating the refrigeration system only will have the same effect.

Air filters

Efficient filters are essential and routine maintenance to clean and replace these is not normally a difficult task with most filter types. It should be a priority periodic maintenance requirement.

There is one exception–the electrostatic type air filter. This type eliminates nearly all dust, dirt and pollen particles from the air by inducing a static electrical charge in the particles as they pass through the filter. Normally air is first passed through a conventional filter to remove most of the impurities. Air is then allowed to pass through the electrostatic filter. After entering the filter the air passes through an ionized field. Electrons and protons which comprise the mass of the atom are equal in number. When an imbalance is created, if the number of electrons is increased, the atom becomes an ion.

A transformer is used to step up the voltage and a rectifier to convert the voltage from AC to DC. This way a very high voltage is generated. Wires with a high positive voltage are located between earth plates. The electrons passing through the air space put a positive electrical charge on any particles of matter entering the 'ionized field'. These particles are then attracted to the earth plates which have a negative potential. Potentials of over 1200 volts can be experienced in this type of filter. This high voltage could present a grave problem for an engineer not conversant with this type of filter.

Important: When it becomes necessary to clean an electro-static filter, under no circumstances should this be attempted whilst the circuit is live. A good safety procedure would be to have a second engineer in attendance when a cleaning operation is being carried out to ensure that the circuit is at all times safe. The filter isolating switch should be turned to the OFF position and remain so while cleaning is being carried out. Some models require the ionized field to be discharged to earth before any work is carried out and includes a facility for this purpose. **No attempt should be made to clean a filter 'in situ' with wet hands until this is completed**. Filters can be cleaned with hot water.

Fire dampers

All ductwork will convey fumes, smoke and flames from areas where there is a fire. Automatic fire dampers are mandatory these days in all ductwork systems. They provide protection against explosions and against circulation of toxic fumes to those areas not affected by a fire. They also prevent any rapid

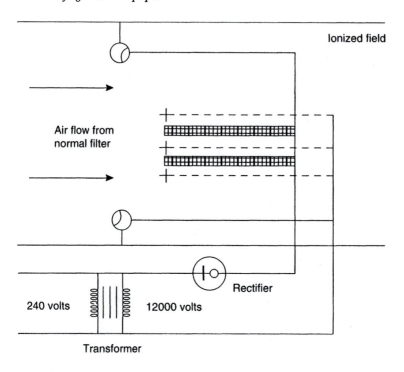

Ionized field

Air flow from
normal filter

Rectifier

240 volts

12000 volts

Transformer

Figure A6   *Electrostatic filter—principles of operation*

spreading of a fire by restricting the air supply, thus depriving any fire with oxygen.

It should be the duty of any service or maintenance engineer to ensure that all dampers are operational after carrying out work on an air conditioning system.

Air conditioning systems

There are many different types of air conditioning systems, each catering for specific requirement. These can be placed into four separate groups:

1   Ventilating
2   Packaged units
3   Window units
4   Split systems.

Ventilating units

Ventilating units are mainly used in the commercial and industrial field for the purpose of removing fumes and polluted air from processing areas. Examples of these would be paint spraying, welding and saw mills. These working areas would require a plant capable of removing and supplying large quantities of fresh air to dilute the polluted air within the work space.

These systems can operate on the exhaust method by removing the polluted air which lowers the pressure in the work area so that fresh air can easily be drawn in to take its place. Another method is to supply air under pressure from fans to create a pressure build-up in the work area. This pressure will then exhaust the polluted air. It is also possible to have a combination of both methods. These installations operate without cooling or heating. In areas where toxic fumes present health risks, safety regulations specify that duplicate fans should be installed in case of failure and at least one fan should operate at all times when personnel are working.

Packaged units

Packaged units embrace a host of different designs and sizes which are constructed so that the air handling and refrigeration circuits are totally enclosed in a cabinet type housing. Those designs with air cooled condensers are normally small capacity and operate on single phase electrical supply. They are usually factory charged with refrigerant. Larger capacity models will require three phase supply and possibly some amount of ductwork. The capacity range of these units can be as low as 400 W up to 20 kW for the larger units. It is important that these units are located in places where the ambient temperature or supply of fresh air temperature is not too high at peak periods. It is not advisable to locate them in a position where they will be exposed to long periods of direct sunlight or in places where the intake of fresh air is restricted. High condensing temperatures will produce higher evaporating temperatures so that space conditions will not be easily maintained.

Window air conditioners

Window air conditioners are ideal for small rooms and offices to supply both cooling and heating as required. They operate on a 'reverse cycle' principle which is manually selected on most models. The units are designed with totally enclosed evaporator/condenser coils in a compact housing. The unit can be installed in a wall or window frame with the outside coil exposed to outside ambient air. Window conditioners operate satisfactorily with ambient air temperatures in excess of 7.5 °C.

Selecting the unit to provide heating or cooling alters the refrigerant flow so that the indoor coil of the unit acts as a condenser and the outdoor coil as an evaporator. When refrigerant flow is controlled by a capillary, routeing the suction and discharge of the compressor to relevant coils is simple because refrigerant will flow through a capillary in either direction. A more detailed description of a reversing valve system has already been given in Chapter 11.

Split systems

Split systems consist of two separate units, the indoor air handling unit which may be located within a conditioned space and the outdoor unit which houses a cooling coil, fan or fans and a condensing unit.

The outdoor unit must be located where it can be supplied with sufficient quantities of condensing fresh air which can then be discharged without affecting the conditioned space. This type of system will require installation of refrigerant pipework, a certain amount of ductwork and electrical supply to both units.

The split system will also require leak testing, evacuation and charging with refrigerant like a normal refrigeration system. Reverse cycle operation can be achieved by installing a reversing valve in the outdoor housing close to the compressor. In the interest of safety a main isolator switch should be located inside the conditioned space adjacent to the indoor unit. The systems operate on thermostat control.

Note: These systems and some of the packaged units will require condensate drain facilities and if the drain outlet terminates at a rain drain or sewer the condensate tubing must incorporate a trap at the outlet. This will prevent undesirable odours infiltrating into the conditioned space.

All reversing valve systems incorporate a reverse flow heat pump filter drier as this type of filter drier allows refrigerant to flow through the filter core in either direction. This prevents foreign bodies from being released into the system when reversal of the refrigerant flow takes place. These types of filters are very efficient, capable of maximum moisture retention and resistant to reactive chemical substances which may be circulating in the refrigerant and oil.

Appendix B   Refrigerant data

The refrigerants most commonly used in domestic, commercial and industrial systems are the following:

R11	Employed with centrifugal compressors for air conditioning systems, as a secondary refrigerant and also as a solvent
R12	Until recently the most widely used for high, medium and low temperature applications
R13	Used for ultra-low temperature applications
R22	Used in commercial and industrial low temperature systems and in some domestic appliances
R113	Used mainly in comfort air conditioning systems with centrifugal compressors
R500	Azeotropic mixture of R12 (73.8%) and R152 (26.2%) – substitution for R12 can increase compressor capacity
R502	Azeotropic mixture of R22 (48.8%) and R115 (51.2%) – developed for low temperature applications to replace R22
R717 ammonia	Has higher refrigerating effect per unit capacity than any other refrigerant – used in large industrial systems

The installation or service engineer should be able to identify refrigerants by the cylinder base colour and associated label colour, or by the cylinder base colour and the colours of the bands located around the top of the cylinder. The cylinder colour codes for the refrigerants listed above are given in Table A1.

Table B1   *Refrigerant cylinder colour coding*

Refrigerant	Cylinder base	Band(s) or label
R11	Grey or silver	Purple/orange label
R12	Grey or silver	Purple/light blue label
R13	Grey or silver	Purple/yellow label
R22	Grey or silver	Purple/green label
R113	Grey or silver	Purple/dark blue label
R500	Grey or silver	Purple/grey label
R502	Grey or silver	Purple/brown label
R717 ammonia	Black	Red and yellow bands

Refrigerant pressure/temperature charts

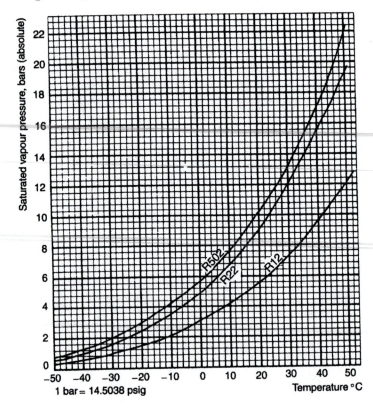

1 bar = 14.5038 psig

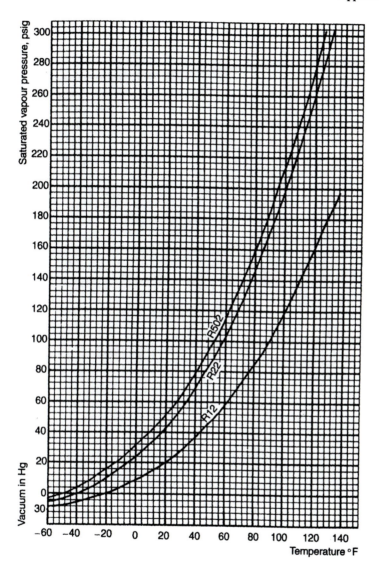

Saturated vapour pressure of refrigerants

Temperature °C	Absolute pressure, bar					
	R11	R12	R13	R22	R502	R717
−60		0.226	2.799	0.378	0.433	
−55		0.300	3.438	0.498	0.637	
−50	0.027	0.392	4.182	0.647	0.820	0.409
−45	0.037	0.504	5.039	0.830	1.042	0.545
−40	0.051	0.642	6.022	1.052	1.309	0.718
−35	0.069	0.807	7.139	1.318	1.626	0.932
−30	0.093	1.004	8.403	1.635	1.998	1.196
−25	0.122	1.237	9.825	2.088	2.433	1.516
−20	0.158	1.509	11.416	2.444	2.937	1.902
−15	0.203	1.826	13.188	2.951	3.515	2.365
−10	0.257	2.191	15.154	3.534	4.174	2.908
−5	0.323	2.610	17.327	4.202	4.922	3.552
0	0.402	3.086	19.723	4.963	5.764	4.295
5	0.495	3.626	22.356	5.823	6.708	5.161
10	0.605	4.233	25.243	6.792	7.762	6.149
15	0.734	4.914	28.406	7.877	8.932	7.287
20	0.883	5.673	31.871	9.087	10.226	8.570
25	1.055	6.516	35.675	10.432	11.652	10.010
30	1.252	7.449	critical	11.921	13.219	11.670
35	1.476	8.477		13.562	14.936	13.500
40	1.731	9.607		15.365	16.812	15.540
45	2.018	10.840		17.341	18.859	17.820
50	2.340	12.190		19.499	21.087	20.330

Standard atmosphere = 1.01325 bar

R11	trichloromonofluoromethane
R12	dichlorodifluoromethane
R13	monochlorotrifluoromethane
R22	monochlorodifluoromethane
R502	azeotropic (R22 + R115)
R717	organic NH_3

Index

Absorption systems, 101–3
Accumulators, 91–2, 94
Acid testing, 61
Aerosols, 176–7
Air, 11, 195–205
  velocity, 197, 203–4
Air conditioning, 131, 194–214
  air filters, 211
  ductwork, 209–10
  fans, 209–10
  maintenance, 209–12
  noise, 209–10
Air cooled systems, 55–6
Air filters, air conditioning, 211
Air motion, 202–4
Air movement, 202–5
Air temperature, 194–5, 198–202, 205–7
Algaecides, 60
Ammonia, 19–20, 101–2, 132, 169–70, 215–7
Appliance systems, 90–2
Aqua-ammonia, 101–3
Atmosphere, 195–6
Axial fans, 209
Azeotropic blends, 25, 181

Bearings, 153
Bi-metal switches, 83
Bimetal thermostats, 69–71
Blowtorches, 51, 126–7
Brazing, 124–7
Bubble point, 181, 191
Bubble test, 18, 21
Burn-out driers, 62–3, 135–5
Burn-out test indicator, 61

Calipers, 144–5
Capacitors, 80–2, 88–9
Capillary restrictors, 91, 93
Carbon, 196
Carbon dioxide, 175–6, 196
Castrol, 187
Ceiling mounted condensing units, 115, 117
Centrifugal fans, 209
Centrifugal switches, 82–3

CFCs, 169–72, 175–9, 185–6
Charging, see Liquid charging; Vapour
  charging
Check valves, 134–5
Chloroflourocarbons, see CFCs
Circuit breakers, 83, 150–1
  magnetic, 150–1
  thermal, 150
Colour coding, 9, 215–16
Comfort conditions, 204–7
Commissioning, 159–72
Commissioning checks, 163–4
Compound bar thermostats, see Bimetal
  thermostats
Compound gauges, 9
Compressor charging, 167
Compressor efficiency test, 51
Compressor oils, 187
  draining, 168
Compressor pressures, 37–8, 72–3
Compressor pump test, see Compressor
  efficiency test
Compressor rotary shaft seals, 53
Compressor valve plate assembly, 52–3
Compressors, 2, 41–2, 74–5, 90, 104, 107–8
  hermetic, 7, 12–13, 26
  open-type, 7, 11–12, 36–7
  semi-hermetic, 7, 26, 35
Condensers, 4
  cleaning, 170
Condensing medium, temperature, 35–7
Contactors, 155–6
Contamination, 11, 17, 60, 90, 121, 159–60
Contractors, 159
Cooling coil, see Evaporators
Copper plating, 159
Copper tubing, 124–7
Corrosion, 60
Crankcase equalizing, 121–3
Crankcase heaters, 133–4
Crankcase pressure regulators, 128–9
Current overload, see Electrical circuits,
  protection
Current relay, 75–6

Decontamination, 97–8
Deep evacuation method, 170
Deep vacuum method, 160–1
Descaling, see Scaling
Dew point, 181, 199
Diagnosis, 1, 31, 34, 41–73
    electrical faults, 96–7
    vapour compression systems, 104–8
Dial-a-charge, see Visual charging cylinders
Dilution method, 161, 170
Direct drive units, 55
Discharge line equalizing, 121–2
Discharge line oil traps, 117–18
Discharge lines, 121–2
Discharge mufflers, 120
Discharge reeds, 39–40
Discharge service valves, 7
Distributor refrigerant controls, 46
DOL starters, 157–8
Domestic absorption system, 101–3
Domestic systems, 90–103, 189–90
Drive belts, 139–41
Drive couplings, 2, 142–6
Drive motors, 152–8
    connections, 154–5
Drive pulley size, 72
Dry bulb temperature, 198, 201
Ductwork, air conditioning, 209–10
Dye systems, 18

Earthing, 152–5
Effective temperature, 205–6
Electrical circuits, 96–8
    protection, 147–51
Electrical faults, 96–7
Electrical test procedures, 84–9
Electromagnetic valves, 131
Electronic leak detectors, 19
Electrostatic air filters, 211–12
Engineers, 1–2
Environment, 169
Equipment, 126–7
Evacuation, 63, 160–3
Evaporating temperature, 33–4
Evaporator pressure regulators, 129
Evaporators, 38–9, 41, 43–5
Excessive operating head pressure, 55–7
Expansion devices, 4
Expansion valve capacity, 106
Expansion valves, 39, 41–2, 44–5, 47, 72–3
    replacing, 46–9
    setting, 43, 184
External equalizing, 43–5
External equalizing capillary, 49

Fans, air conditioning, 209–10
Fault finding, see Diagnosis
Fill ratios, 29–30
Filter driers, 38, 135–5, 214
    replacing, 49–51
Fire dampers, 211–12
Flare fittings, 124
Flare nuts, 112–13
Flare unions, 112–13
Flexible couplings, 114–15
Food products, 31–2
Freezer door heaters, 94
Freezing, 160
FREON refrigerants, 180, 192
Fuses, 83, 147–50
    cartridge, 148
    HRC, 148–9
    rewireable, 147–8
Fusible plugs, 131–2
Fusing factor, 149–50

Gas refrigerators, 101–3
Gauge manifolds, 1
Gauges, 1
    fitting, 11–13
    removing, 13–14
Glide, 181–4
Global warming, 175–6
Greenhouse effect, 169, 175–7
Greenhouse gases, 175–6

Halide torches, 18, 21
Halogenated refrigerants, 20
HCFC22, 177
HCFCs, 176, 178–80, 185
HFCs, 176, 178–80
High side purging, 57–8
High suction, 39–40
Hoses, 9–14
Hot gas defrosting, 94–5
Hot wire anemometers, 203
Humidistats, 202
Humidity, 34, 194–5, 199, 201–2, 206–7
Humidity control, 202
Hydrochloroflourocarbons, see HCFCs
Hydrofene, 60
Hydroflourocarbons, see HFCs
Hydrogen, 196
Hygrometers, 201
Hygroscopic ester oils, 25
Hydra-zorb, 113–15

IC-22 relay, 80–1
ICI, 178

Installation procedures, 2
Invar, 69

Klea refrigerants, 178–9, 184, 187

Leak tests, *see* Refrigerant leaks, detection
Level indicator gauges, 143–4
Line tap valves, 15–16
Liquid charging, 27–9
Liquid charging valves, 47
Liquid flushing, 97
Liquid knock, 106–7
Liquid receivers, 4
Liquid shut-off valves, 7
Litmus paper, 61
Low discharge pressures, 39–40
Low pressure cut-out point, 33–4
Low side purging, 50
Lubricating oils, 6
Lubrication, 152

Magnetic valves, 39
Methane, 175–6
Moisture, 50–1, 90
Monometers, 204–5
Montreal Protocol, 169, 175–7
Motor burn-out, 60–3
Motor cycling controls, 69–71
Motor protection, 83–4
Motor windings, 75
Munsen ring, 113–15

Negative temperature coefficient devices, 84
Nessler's reagent, 19
Nitrogen, 196
Noise, 71–2, 105
    air conditioning, 209–10
Non-return valves, *see* Check valves

OFN, *see* Oxygen-free nitrogen
Oil, 40, 168, 184–7
    adding, 165–7
    replacement, 54–5
Oil charging pumps, 165–6
Oil cooling, 94
Oil levels, 164–9
Oil pressure, 136–7
Oil pressure failure switches, 136–8
Oil separators, 118–20
Oil traps, 2, 117–23
Open drive units, 53–4, 82–3
Open-type systems, 11
Operating head pressure, 34–7
Operating principles, 31–40

Overload protectors, 74–5, 83–5
Oxygen, 196
Oxygen-free nitrogen, 20–1, 161
Ozone depletion, 175–7
Ozone depletion potential, 177–8
Ozone layer, 175

Packaged units, 213
Parallel pipework, 121–3
Phosgene, 60–1
Piercing valves, *see* Line tap valves
Pipework, 2, 111–27 ·
    assembly, 124–7
    installation, 111–16
Pipework fittings, 112–13
Pipework routes, 115–17
Pipework supports, 113–15
Pitot tubes, 203–4
Pollution, 2, 13–14, 58, 169
Positive temperature coefficient devices,
        78–80, 84
Potential relay, 77–8, 89
Pressure controls, 14, 64–8
    setting procedures, 65–8
Pressure drop, 49
Pressure gauges, 7–15
Pressure testing, 20–1
Pressure/temperature relationship, 181–3,
        191–3
Pressures, 1, 7–11, 38–40, 43–5, 105–6,
        128–32
    recording, 13
Psychometric charts, 197, 199–202
Psychometry, 197–202
PTC, *see* Positive temperature coefficient
        device
Pumping down, 24–5

R11, 62, 97, 177, 215–17
R12, 14, 29, 32, 44, 60–1, 99, 132, 177, 181–2,
        215–17
R13, 215–17
R22, 14, 21, 29, 60–1, 99, 132, 160, 178,
        181–2, 184, 215–17
R113, 177, 215–16
R114, 177
R115, 177
R134a, 25, 186
R404, 181
R404A, 181, 183
R407A, 181, 183
R500, 160, 177, 215–17
R502, 14, 25, 29, 60–1, 99, 132, 177, 181–2,
        184, 215–17
R717, 132, 215–17

Radial fans, 209
Rapid chargers, 27
Receiver valves, 9
Refrigerant blends, 181–4
Refrigerant charges, 29, 101
   changing, 184–7
Refrigerant charging, 29, 97, 99–101, 163
Refrigerant cylinders, 29–30, 61, 170–1
Refrigerant hoses, 188–90
Refrigerant leaks, 1, 41, 90
   detection, 17–23, 113, 163, 177
Refrigerant metering controls, *see* Expansion
   devices
Refrigerant pressure/temperature charts,
   181–3, 191–3, 216–17
Refrigerant recovery systems, 171–2, 185,
   187–8
Refrigerant thermal decomposition, 60–1
Refrigerants, 4, 159, 176, 215–16
   adding, 25
   azeotropic, 25, 181
   glide, 181–4
   halogenated, 20
   handling, 11, 21–3, 170–1
   loss of, 13–14
   new, 29, 175, 178–93
   phasing-out, 6
   reclaiming, 185, 187–8
   removal, 47–8, 170–1
   shortage, 34, 38–9
   types, 72–3
   vapour pressure, 218
   zeotropic, 25, 181
Refrigeration circuits, 90–2
Refrigeration vapour compression systems,
   177
Relative humidity, *see* Humidity
Remote drive motors, 107–8
Reseating, 52
Reversing valves, 131, 213–14
Rotating vane anemometers, 203
Rotation, 153

Safety precautions, 11, 21–3, 58, 60, 125, 159,
   170–1, 211
   see also Refrigerants, handling
Scaling, 56–60
Schraeder valves, 7, 14–15, 188
Sealed systems, 71, 170–1, 189–90
   see also Domestic systems
Service calls, 1–2
Service equipment, 72, 75
Service gauge manifolds, 8–13
Service gauges, 32

Service valves, 1, 4, 7–8
Shims, 145–6, 153
Sight glasses, 38, 99, 134–5, 165–6
Silver soldering, *see* Brazing
Single phase/three phase compressors, 107–8
Sling psychrometers, 201
Sludging, 160
Solenoid flow controls, 94–6
Solenoid valves, 39, 133
Split systems, 214
Start capacitors, 75, 77
Starters, 74, 77–80, 155–8
   direct-on-line, 157–8
Static pressure, 204
Suction line accumulators, 91–2, 94
Suction line equalizing, 123
Suction line frosting, 45–6
Suction line oil traps, 117–18
Suction reeds, 39–40
Suction service valves, 7
Suction vapour cooling, 26–7
Sulphur candles, 19–20
Supply and control switch, 84–5
SUVA refrigerants, 179, 193
System analysers, *see* Service gauge manifolds
System control valves, 128–32
System evacuation, 2
System faults, 71
   symptoms, 38–40
System flushing, 62–3
System pressure method, 168
System service valves, *see* Service valves

TD, *see* Temperature difference
Temperature, 31–7, 41–3, 72, 104
   air, 194–5, 198–202, 205–7
Temperature difference, 31–4, 41–2, 72–3
Terminal board connections, 155
Test cords, 86–8
Test pressures, 17–18
Test procedures, 84–9
Thermal bulbs, 41–2, 44
Thermal charge, loss of, 47–8
Thermal overloads, 153
Thermal shock, 206–7
Thermistors, 83–4
Thermostatic expansion valves, *see* Expansion
   valves
Thermostats, 69–71
Tools, *see* Service equipment
Trace pressure tests, 21
Triple evacuation method, *see* Dilution
   method
Tubing saddles, 113–5

Ultra violet light, 176

Vacuum method, 168
Vacuum pumps, 162–3, 166–7
Valve reeds, 39–40
Valve settings, 43
Vapour charging, 25–7
Vapour compression systems, 2–4, 104–8
Vapour pressure, 218
Vapour pressure thermostats, 69
Vapourization, 181–4
Velocity pressure, 204
Velometers, 203
Ventilating units, 213
Ventilation, 21, 23
Vibration loops, 112–13
Visual charging cylinders, 98–100

Voltage relay, *see* Potential relay

Water coils, 35–6
Water cooled condensers, 29, 58–60
Water cooled systems, 56–7
Water pressure/temperature, 163–4
Water regulating valves, 129–30
Water temperature, 37
Water vapour, 195–6
Wattmeter, 75
Wet bulb depression, 201
Wet bulb temperature, 198–201
Windings, 85–6
Window air conditioners, 213

Zeotropic blends, 181

9 780750 636889

Printed in the United Kingdom
by Lightning Source UK Ltd.
108843UKS00003B/64-408